WILDERNESS SECRETS REVEALED

WILDERNESS SECRETS REVEALED

ADVENTURES OF A SURVIVOR

ANDRÉ-FRANÇOIS BOURBEAU
Author of *Surviethon*

Foreword by Les Stroud

DUNDURN
TORONTO

Copyright © André-Françpis Bourbeau, 2013

All rights reserved. No part of this publication may be reproduced, stored in a retrieval system, or transmitted in any form or by any means, electronic, mechanical, photocopying, recording, or otherwise (except for brief passages for purposes of review) without the prior permission of Dundurn Press. Permission to photocopy should be requested from Access Copyright.

Editor: Jennifer McKnight
Design: Jennifer Scott
Printer: Webcom

Library and Archives Canada Cataloguing in Publication

Bourbeau, André-François, 1953-
    Wilderness secrets revealed : adventures of a survivor / André-François Bourbeau.

Also issued in electronic format.
ISBN 978-1-4597-0696-5

1. Wilderness survival. 2. Bourbeau, André-François, 1953-. I. Title.

GV200.5.B687 2013   613.6'9   C2012-908623-1

1   2   3   4   5   17   16   15   14   13

We acknowledge the support of the **Canada Council for the Arts** and the **Ontario Arts Council** for our publishing program. We also acknowledge the financial support of the **Government of Canada** through the **Canada Book Fund** and **Livres Canada Books**, and the Government of Ontario through the **Ontario Book Publishing Tax Credit** and the **Ontario Media Development Corporation**.

Care has been taken to trace the ownership of copyright material used in this book. The author and the publisher welcome any information enabling them to rectify any references or credits in subsequent editions.

J. Kirk Howard, President

Printed and bound in Canada.

All images, unless otherwise stated, are from the collection of the author.
Front cover: The author at sixty years young, playing a game of "lost in the woods." *Photo by Michel Martineau.*
Back cover: The author searching for firewood in the middle of the night.

Visit us at
Dundurn.com | Definingcanada.ca | @dundurnpress | Facebook.com/dundurnpress

Dundurn	Gazelle Book Services Limited	Dundurn
3 Church Street, Suite 500	White Cross Mills	2250 Military Road
Toronto, Ontario, Canada	High Town, Lancaster, England	Tonawanda, NY
M5E 1M2	LA1 4XS	U.S.A. 14150

*To my dear mom, Gertrude,
who was worried more than once*

Table of Contents

Foreword by Les Stroud		9
Acknowledgements		13
Introduction		17
One	Ready, Set, Go!	19
Two	The Hunt	23
Three	The Rug	27
Four	Spicy	31
Five	The Swamp Tramp	37
Six	Up Here, Sir!	47
Seven	A Sheep	55
Eight	The Farmer	61
Nine	Sandals	73
Ten	Licking My Chops	79
Eleven	Batman	83
Twelve	Beans	89
Thirteen	Banquet Time	97
Fourteen	Batman Again	107
Fifteen	A Cold Hole	117
Sixteen	Ferns Crack Me Up	123
Seventeen	Exquisite Rain	133

Eighteen	The Charcoaled Commandos	143
Nineteen	Happy Birthday	167
Twenty	The Philosopher	183
Twenty-One	Retro-Propulsion	203
Twenty-Two	The Birchbark Pillow	223
Twenty-Three	The Opera Singer	237
Conclusion		251
Favourite Books		261
About the Author		263

Foreword

André-François Bourbeau is obsessed with survival and adventure. Some would say passionate. I say obsessed!

Until I endeavoured to put *Survivorman* on prime-time TV, the practice of survival techniques was relegated to the odd local TV station feel-good story. Those of us involved in survival learning were — well, let's face it — survival geeks. We lived and breathed fire-making methods, shelter construction, signaling methods, primitive earth skills, and, my personal favorite, edible wild plants. We were usually a ragtag bunch of mostly men that either had fond memories of our Boy Scout experiences or considered *Jeremiah Johnson*, hands down, our favorite movie of all time. We wanted to play mountain man. We were taken out into the bush and taught by a small handful of teachers throughout North America. They were nature nuts, survival nuts, bush nuts, and often gear geeks. But man they sure knew how to start a fire by rubbing two sticks together and follow it up with a bowl crafted of birchbark. Those of us in the know are aware of exactly who most of them are. They would show up at primitive skills rendezvous in legend status. In those circumstances the name of André-Francois Bourbeau was not legendary. Hell, it was barely known, because he was evolving in Quebec, mostly alone. Yet he was and is deserving of legend status. In my opinion, he is quite possibly the best there is. And why is that? Because he is freakin' obsessed!

I first met André somewhere around 1993 or '94 when my then wife and I were gearing up to spend a year living in the wilderness as if it were five hundred years ago. No metal, no matches, no nylon tent. Just us, birchbark, and maple syrup (well, more or less). In Quebec he had been able to take his love for survival skills into a much more prominent place of media attention with his Survivathon ordeal. And it was this experience in the field I was after for advice on what we were about to attempt.

Open and welcoming to us from the start, he LOVED the opportunity to help us out with advice. Did I mention he is obsessed? On our first meeting he started talking fire-starting. Well let me back up a little. I should admit I absolutely adore survival skills of all kinds. Modern, pioneer, primitive earth skills, you name it. But I am decidedly, believe it or not, NOT obsessed. So it should come as no surprise that my eyes started to gloss over as André went into an intense discussion about how to start a fire using a simple stainless steel tablespoon by reflecting the sun with the concave part. On one occasion, many hours of endless talk of survival methods and adventures later, he surprised me. He pulled out his guitar and told me he had decided to learn how to play. He wouldn't have, except I mentioned I was a musician and, well, that was all he needed. He then endeavoured to play for us his rendition of "Achy Breaky Heart" with a thick French-Canadian accent while we sat awkwardly on his couch pretending to appreciate the assault on our senses. (His ability at pitch, technical playing ability, and timing were a universe away from his outdoor skills — sorry, my friend.) But. He was so, I mean *so* into it. How could he not be? He is an intense man with a passion as big as the mountains. And his passion for survival skills is only outdone by his passion for life.

I am glad that he started this book with a quote from Stefansson, the famous Arctic explorer. Because Bourbeau, like Stefansson and like me, suffer the same fate. We know too much. We *can't* get lost. We are *always* prepared. And I can tell you with great certainty that every hardcore survival instructor out there secretly harbours a desire to blow it just once, thoroughly and pathetically, enough so that finally we can *really* put our skills to the test. No students. No pick up on Sunday afternoon. No backpack of survival gear. Put to the test, fully and completely. Just like, well,

Foreword

just like all the stories guys like André and I have been reading all our lives. Heroes à la Douglas Mawson, the Dougal Robertson family, and the survivors of the famous Andes plane crash.

But we can't. André can't. So. We volunteer. André with his Survivathon and a hundred other expeditions and me with my *Survivorman* ordeals. We try to come as close as we can to really putting our skills to the test. And we fail, we know, every single time. Because we aren't in a real plane crash and we didn't really lose our canoe in the Arctic and we aren't really lost in the jungle. But we obsess about it. And André, well, when it comes to wilderness adventure and survival, he is the greatest and most obsessed of us all. When it comes to survival, he is legendary.

Les Stroud
Author, producer, and star of *Survivorman*.

Acknowledgements

My dear friends and family,

Bob Henderson, mon ami, it's your unbounded enthusiasm that convinced me to initiate this book project. It's so inspiring to share history with you while canoe tripping in the wildest reaches of Canada. Amanda Gibson, my precious first reader, you paddled through the mire of this book with me with such perfect complicity. How wonderful it was to share this project with you! Love you so much. Billy Rioux and Jean-Charles Fortin, my trustworthy and dear friends, I so appreciated you reading my first manuscript to give me your precious advice.

Les Stroud, I've witnessed your gusto from the very beginning of your Survivorman journey and have greatly admired all you have accomplished since. What an outstanding career. Thank you so kindly for having accepted to write the foreword to this book.

Jacques Montminy, I could not have chosen a better partner for the Survivathon adventure. I never told you, but your calm inner strength really helped me overcome. My dear brother Michel, my life would not have been the same without your quirky sense of humour. You are a unique breed indeed. Gaëtan, Guy, Réal, my other dear brothers, I could always count on you three to participate in adventures and laugh at Michel and me. You are part of my very best memories. Hey, Rob Bicevkis, pal of so many of my early adventures, we must do more now that we've both retired. Your intelligence and wit have always inspired me. James Déraps, Michel Martineau, and Marcel Savoie, my Retro-Propulsion partners and long time friends, you have always encouraged my folly. Never a dull moment with you guys around. Mario Bilodeau, dear brother-like colleague, your presence at my side for thirty years comforted me so. Our new fishing career is just beginning. And then there are all of you, former members of the Thornlea High School Outing Club where I grew up. I am so happy when I receive news of you guys and gals. We must organize a family reunion trip and invite Angus Baptiste!

Jimmy Bossum and Gérard Siméon, my Native American mentors, memory of you will forever be encrusted in my brain. You were the end of an era, but your skills, talent, and wisdom live on. Same with you Kirk Wipper and William E. Harmon; joy still flows when I remember you. You were my second dads. And Doctor Taylor Statten, I will forever be indebted to you for guiding me so preciously over all of these years. An inspiration in the greatest sense.

I learned myriad technical tricks and gained countless insights from you Mike Obarymsky, Harland Gold Metcalf, Alexandra Conover-Bennett, Garrett Conover, Ron Hood, Barry Keegan, Chuck Chase, Steve Watts, Benjamin Pressley, Denis Morissett, and George Hedgepeth. And thanks for experimenting at my side Manu Tranquard, Frédéric Fournier, Pierre Bossé, Gilles Levesque, Jeff Thuot, Martin Gagnon, Frédéric Dion, Frédéric Parent, Marc-Olivier Forget, Jean-François Dubé, David Boulais, Nicolas Letourneau, and Jesse Schobb. May we continue to share superb moments.

Barry Penhale, publisher emeritus, and Jane Gibson, editor, you believed in me. It was precious to feel the support of such an experienced team. I also appreciated working with the fine group at Dundurn, real professionals. Hats off to you Jennifer McKnight. What editing skills! Bravo. Nice to have crossed your paths Shannon Whibbs, Sheila Douglas, Caitlyn Stewart, Karen McMullin, James Hatch, and Margaret Bryant.

As for you, dear university students, there are so very many of you that I have grown extremely fond of. I won't name all of you here, but you know who you are, because you visit me at home. I have gained so much sharing with all of you.

Acknowledgements

Dad, your true passion lit up my life. You were my coach, my mentor, my buddy, my confidant. You live on through me. Mom, I dedicate this book to you. I hope that says it all, because there are no words strong enough to recompense you. Danièle and Renée, my dear sisters, you were wise not to follow in my adventures. But I always knew you were there with me in spirit — it was obvious. Kisses to you both.

My other cherished friends: Johanne, Richard, Jean-Claude, Cécile, Lise, Vincent, Eve, Eddy, France, Marc, Gilles, Diane, Linda, Cylvie, Mario, Véronique, Jacques, and Martina. Your support throughout all of this has been priceless.

And for dessert there are you two treasures, Lizon and Veronica, my loving family. You make my life worth living each and every day.

To all of you, I express my most sincere and heartfelt thanks. May your lives be long and serene.

André-François

Introduction

"My favorite thesis is that an adventure is a sign of incompetence. If everything is well managed, if there are no miscalculations or mistakes, then the things that happen are only the things you expected to happen, and for which you are ready and with which you can therefore deal. Being thoroughly alive to the truth of this principle, I am also thoroughly ashamed of owning up to such adventures as we have had."

— Vilhjalmur Stefansson

If Stefansson's postulation from his 1913 *My Life with the Eskimo* is right, and my own experience indicates it surely seems to be so, well I too must fess up to being thoroughly ashamed. For I have encountered my good share of adventures, and therefore I freely admit to having been most incompetent! The only redeeming factor, which comforts me, is that the last twenty years have been fairly monotonous. I suppose all of the close calls of my first four decades, and the wisdom thus gained, naturally culminated in a slightly droning and adventure-deficient outdoor education career. But my job at that point was to let others live the excitement by bringing them into what Jean Brunelle calls their "zone of delicious uncertainty," albeit in a controlled setting. In this I have been mostly successful, thankfully; neither I nor any of my students have suffered a serious mishap since. How boring.

Indeed, life was more exciting when I dared let my students rappel down cliffs on our homemade grass ropes. Even if I tested the rope by going first, as I always did, I would today be crucified in a public place for such an act. Risk management has overcome our world, and I myself am responsible for this to a certain extent, having written Quebec's outdoor risk management reference manual. Saving lives and preventing accidents is good, I suppose, but I still pine for a canoe trip without maps, where you get hopelessly lost while looking for a non-existent portage to a non-existent lake. I still long for the right to plan a trip poorly, secretly yearning to scavenge cattail roots and wild mushrooms for a few days.

Well, this book will send us back into an era when risks were less scrutinized. Yes, Mr. Stefansson, I am guilty as charged of incompetence. At least I had fun.

An unusual rappel.

One

Ready, Set, Go!

> *"Success isn't a result of spontaneous combustion. You must set yourself on fire."*
> — Arnold H. Glasow

I was elated to have discovered in Jacques a trustworthy partner for an outlandish month-long stint in the woods. We shook hands for an August 1 departure.

Ready…

Once the trip was confirmed I could concentrate on the logistical details. Finding an area devoid of human habitation turned out to be a challenge. It was finally through the Ministry of Natural Resources that a sufficiently large tract of virgin Crown land was identified near the 52nd parallel. Blindfolded, I tossed a dart on the provided map. The exact drop off spot was kept secret from me. It didn't really matter.

Next I concocted a research methodology in order to transform the trip into a scientific experiment. We would partake in anthropometric and fitness measurements before, after, and two months post-trip. Same with complete blood sample analysis. I would also perform a task-versus-time analysis using portable voice recorders. Carrying a small weigh scale would permit me to accurately measure our caloric intake of wild plants and animals. Finally, I would record rescue possibilities during the month-long stay by noting the passage of aircrafts.

In June came an unexpected and welcomed surprise. My good friend Jean-Claude Larouche had barely reacted when I had mentioned our survival trip, but a couple of weeks later invited me to lunch. At the time he was working at raising funds for a foundation associated with the famous St-Félicien zoo with the intention of building a "Nature's University" where animals would roam free and the people studying them would be in cages. He sprung the question just before dessert. Would I agree to him using the "Survivathon" to promote the project? Cool name for our adventure. I didn't even blink. Sure!

From that moment on, Jean-Claude amusingly called himself my "agent." He had unwavering faith in my capability to see myself through the one-month ordeal and started promoting the event. Oh, and what a brilliant promoter he turned out to be. In no time he had confirmed the media's interest in the Survivathon and had convinced the local flying school to furnish a helicopter to ceremoniously drop Jacques and me into the woods.

We decided to enter the wilderness dressed as city folk out for an afternoon of berry picking. In downtown Chicoutimi, I stopped the tenth person to walk by me on Rue Racine, the main street. That's how I selected our gear list, by copying that person's possessions: one pair of socks, one pair of leather hiking boots, one pair of jeans, a wallet with two credit cards, a photo of a loved one, $19.65, a ring with five keys, a leather belt with large buckle, a t-shirt, a dress shirt, a light wool sweater, an unlined nylon windbreaker with hood, a bandana, and a felt hat (Jacques went for a baseball cap instead). No matches.

Set ...

Time flew until August 1, when the alarm bell rang to no purpose. It was only six o'clock but I'd already been pacing for an hour, too anxious to sleep, incessantly peering out the foggy window. When Jean-Claude arrived in his ruby red convertible to pick me up half an hour later, I rushed out to sit besides Jacques before he could honk. We headed to a local restaurant, complicity hovering among the three of us.

Delicious and abundant, the breakfast generously provided by the restaurant corresponded entirely to our tastes and needs of the moment.

We wolfed down steaks, eggs, toast, bacon, jam, milk, and fruit, certain we would be dreaming of it all within a few short days. Employees of the restaurant and a few onlookers wished us good luck. We thanked them and took Route 170 to the Bagotville Airport.

A small buzzing crowd welcomed us, the awaiting helicopter creating the soundtrack. With heavy hearts and tears in our eyes, we hugged our best friends warmly, ignoring the journalists vying for our attention. In truth, I was horribly afraid of the trap I'd cornered myself into. I worried about disappointing all of those who believed in us and who had total confidence in our success. As Jacques focused on his family, I had sweet thoughts for dear old mom and dad back in Ontario, feeling like a little boy venturing away from home for the first time.

To the journalist's insistent questions I lied that I was really looking forward to the confrontation with the forest. Deep inside, I knew too well what was awaiting me.

Jean-Claude then proceeded to publicly check the contents of our scientific equipment and our pocket gear. At the time, I couldn't figure out why he had bothered executing such a summary search; he obviously had no hope of revealing any gear I could have cunningly cached had I wished to cheat. Plus he knew beyond a doubt that my scientific mind would never allow such a grotesque infringement on the objectives of the Survivathon. Always the PR wiz, he had simply wanted to satisfy the skeptical audience.

However, I do confess that the idea of hiding a match somewhere had crossed my mind that morning. But I had immediately realized that I would be lying to myself only. Moreover, that gesture would have undermined my morale to the point of jeopardizing our success.

As we were crossing the threshold of the glass doors, a small package was presented to us. There were four tiny oatmeal cookies, meticulously wrapped and decorated with love by Jean-Claude's girls, Mireille and Eva. That sign of tenderness sure warmed our souls.

Waves of hands cheered us on as we approached the helicopter with tight throats. We were four on board: the pilot, a videographer, Jacques, and myself. The motor roared.

Go!

Two

The Hunt

> "It takes courage to grow up and turn out to be who you really are."
>
> — e.e. cummings

The enormous beast hadn't seen me yet. I discreetly slathered a bit more mud on my face, in my hair too. I crawled closer on scraped elbows. Stop. Just stay immobile for a while. Okay, it was feeding again. I crept closer still. As the early morning dew soaked through the knees of my trousers and the sleeves of my jacket, I felt my heart pounding. Patience. I had to wait for my chance.

It was still there, unaware of my stalking presence. I had got to get to the next shrub undetected. Ever so slowly, I inched along like a spying snake. Made it. The magnificent creature looked like it wanted to flee. No, false alarm. Phew! Shaking, I calmed myself and gently pulled an arrow from my quiver. I licked the feathers for luck as I notched the arrow onto the twisted and greased bowstring. The wait for a broadside shot began.

I observed my prey from a bushy hideout. After interminable minutes it finally moved to present its flanks to me. He sure was a big fat one. I drew my bow carefully to full extension and took aim. There's where those hours of target practise would matter. My release was perfect. Bull's eye!

Darn, how did that squirrel get away? My astounded nine-year-old eyes couldn't believe it. I hit it dead on and it ran away as if nothing happened! Now where was my arrow? I finally located it and examined it with satisfaction, especially the feathers that were hard earned by plucking the tail of a sparrow I caught with an old crate pried up with a stick on a string. That bird sure flew crooked without those two feathers, poor thing. My thumb rubbed the bent beer-cap point that encircled the goldenrod shaft — not sharp enough, I guess. Next time maybe I could figure out how to tie a nail to the tip.

So no squirrel; what would I eat then? I thought I'll head over to the swamp to catch a few frogs. After gathering birchbark and twigs to start a fire, eight miniature frog legs got a cooking. Still hungry. Might as well go fishing. Good thing I swiped those two pins from Sis's doll's diaper. Tied to my kite string, they should do the trick. I uncovered a few shy worms from beneath rocks, seized some mushroom-sucking slugs too, and then headed to the nearby lake. Five sunfish and two perch later, I was back at the fire ring to finish my lunch. The leaves looked like good salad. Nah, way too bitter. Let's try these — even worse. I just ate sour grass as usual. I like those lance-shaped leaves; they taste like lemon. Especially good with fish. For dessert, I gathered some of those little blue wild pears. But maybe Mom had some leftovers of her world-famous chocolate cake. Better head home and check it out!

"Hi Mom! Can I have a piece of cake?"

"Where the heck were you for lunch? I looked all over for you!"

"Oh, I had lunch in the forest." Pride surely radiated from my gleaming face.

"What?"

"Caught some frogs and some fish and ate sour grass and wild pears."

I was the king of the world. Tarzan even.

"Don't you get lost!"

"Don't worry Mom, I always hike up the hill, so to come back I just have to go down to hit the road."

"Okay, here's your cake, and after you're finished don't forget to do a good job on your potatoes!"

The Hunt

Every day I had to peel a seventy-five pound bag of potatoes for dad's restaurant. A good job meant peeling the potatoes thinly. Last time I tried a shortcut, Mom made me whittle down all the thick peels and cooked us mashed potatoes. It took me two extra hours!

That's how my life as an adventurer all started. Mind you, I cringe when I now think of all the poisonous plants I could have "tested" as a child. Good thing I hated mushrooms! My main staple on these outings consisted of those "little wild pears," which I recognized many years later as being juneberries or serviceberries, of the genus *Amelanchier spp*. This is actually a most important plant for wilderness survival, since its bushes grow wild in the middle of nowhere and are widely distributed. Of course, the lance-shaped sour grass of my youth is sheep sorrel, *Rumex acetosella*.

At the time my family was settled in the tiny village of Spragge, on the northern shore of Lake Huron; Dad had set up a restaurant there to profit from the mining boom of the sixties. He loved me a lot — maybe too much. To him I was perhaps a bit of a performance monkey. For example, he had enrolled me at school at only four years old, convincing me I could easily keep up to the older kids. I believed him and did.

I was quite mischievous as a youngster, my idea of fun being cutting my siblings' hair very short to "improve their looks" just before the announced family photograph, taking Dad's brand new watch apart with a screwdriver and hammer to figure out what made it tick, or "sharpening" Dad's chef knives on rocks as a surprise after overhearing that he whetted them on a stone (and then hacking the wooden bed posts to test the results). So it's no wonder that, with me as leader of her seven kids, Mom would order us to "go play outside." And I did. I became an expert at making and using a slingshot — the best were gleaned from red tractor inner tubes — and also most enjoyed tossing rocks with a running-shoe-tongue sling such as David used in the Bible. Let's just say that many Goliath telephone posts bore my scars, and the local police owned a huge collection of my confiscated weaponry.

Just as I hit high school my family moved into town, first to Sudbury for a few years, then Toronto. So I lost my precious contact with nature, concentrating instead on other interests. Individual sports caught my fancy, mostly weight lifting, wrestling, judo, karate, and bicycle racing. Although I wasn't an elite athlete, I did perform fairly well, having won local competitions in all of these disciplines. On later adventures, my physical fitness would sure prove to be handy.

At the age of thirteen I developed a great passion for magic. I ate prestidigitation for breakfast, lunch, and dinner. I was convinced that my career would be that of a professional magician, and I attended any and all available conventions. As an apprentice under the likes of the amazing James Randi and the late Derek Dingle, I learned the tricks of the trade. Enough so that "The Great André" (there is no market for a magician with a humble moniker) could pay his way through university by performing at kids' parties.

It wasn't until I entered university at the tender age of sixteen that nature would summon me once more. The click occurred while reading Jules Verne's *The Mysterious Island*. The engineer Cyril Smith in that story profoundly impressed me, for he could solve all problems encountered while away from modern society. I decided to follow in his footsteps and committed to avoiding dependence on stores to fulfill my basic needs. Discovering nature's magic secrets would become my life-long obsession.

Three

The Rug

"André-François Bourbeau! What in the world are you doing down there in the rec room? It smells like smoke all over the house!"

Now I'm in trouble. When Mom yelled out my full name like that, it sure wasn't a good sign. But darn it, I felt like I was just about to obtain my very first fire by rubbing sticks. Can't stop now! The bow was oscillating in my hands at 200 beats a minute, spinning the drill into a blur. Smoke was pouring out. Sweat too, blinding me.

"Don't worry, Mom. I'm just rubbing some sticks together! I'm almost done." I heard footsteps upstairs.... Quick, push harder André. More speed, more pressure on the drill! The familiar wood powder was piling up, but into a more profuse mound than ever before. Mom was coughing as she came down the stairs, her familiar frame visible through the cloud of smoke.

"*Mon Dieu!*"

"Yyyyeeeeeeaaaa! I did it, Mom, I finally did it! Look at this ember! I've been trying for months and I finally got it!"

"*Es-tu fou*? Are you crazy? You just burned a hole in my new expensive carpet! Why didn't you do this outside?"

"But it's raining like cats and dogs out there!"

"*Misère! Mon tapis!* You just ruined my beautiful Persian rug!"

"My first friction fire ever! Mom, do you realize I can now survive anywhere even without matches? Do you know what this means?"

Starting a fire with a bow drill.

"*Doux Jésus* … They won't be able to fix that. You just wait 'til your dad comes home! Why on my carpet, of all places?"

"That was my problem, Mom. On hard ground my fire board kept moving around and it prevented the spark from forming. On the carpet it was just perfect."

My face was glowing more than the ember, bursting with immense pride. Finally I had attained true survivor status. I was free of society. I could now survive without. Fire, the key to wilderness living. The ultimate symbol of independence. And it was mine, all mine! For free. In my mind, I was reliving prehistory — I was the first caveman to have obtained fire. Nature had let me become her friend. As I stared into the bright coal, in the space of an instant everything around me became fuzzy. I understood, and I was in awe.

"My first friction fire!"

"My rug!"

The Rug

It's been over forty years since that event changed my life, giving me faith and confidence in my capability to surmount any and all obstacles. Back then, you see, there was no Internet and no one to teach me. I had found a small blurb on the bow-drill method of fire-starting in an old book, but there was no indication that a notch was necessary in the edge of the board to oxygenate the charred wood dust. No wonder I had tried unsuccessfully for months. With this practice, my technique had all the while improved. So much so that I finally started my first fire without a notch at all, by drilling so close to the edge of the board that the burnt wood powder simply flowed over the side, forming a pile in which the ember formed. As I told all of my friends about my accomplishment, an old Hungarian granddad (no doubt a real woodsman in his prime) overheard the conversation and smilingly told me about the notch. From then on, starting a fire by rubbing sticks became child's play to me. Or so it seemed after a few dozen successful attempts.

Oh yes, and forty-five years later I guess a serious apology is long overdue.

So very sorry about your carpet Mom! And a special thanks for that knowing smirk, which always shone through the scolding to let me know you were secretly amused.

Four

Spicy

"*Variety is the very spice of life, that gives it all its flavor.*"
— William Cowper

I barged into the house, bellowing with enthusiasm. Dad, having become a master chef and teacher, was in the kitchen as usual, experimenting with new recipe ideas.

"Hey Dad! Check this out. I found a great big patch of wild leeks in the ravine! Let's make some old fashioned *Soupe Bonne Femme*. I'll start peeling the potatoes."

"What? Let me see those."

Dad's eyes lit up; he was always excited by anything to do with food. Already I could sense he was dreaming of new fame with a wild onion soup recipe. He picked up the wild leeks and gave them a sniff.

"These smell pretty garlicky! Are you sure they're leeks?"

"*Wild* leeks, Dad, *Allium tricoccum*. Here's a picture of them in this book on edible plants. Obviously, we'll have to tone down the amount; wild plants are always stronger than the store-bought varieties."

"Let me see that book. Yeah, they kinda look like the picture. Ok, let's cook them up. It's almost suppertime anyway."

Dad pulled a slab of bacon from the fridge and started dicing it up as I peeled a couple of potatoes at lightning speed — I'm the world champ at that task! While Dad cubed the potatoes, I washed the wild leeks and started chopping them up. Tears were streaming down my face.

We tossed the leeks and bacon into the heavy enamelled cast-iron skillet and the familiar onion-bacon smell wafted over the kitchen, but with more oomph than usual. After adding chicken broth and the potatoes, we waited for them to become tender and Dad finished off the recipe with heavy cream. Just in time, Mom walked in with her bags of purchases, followed by the troops, my six younger siblings, who all jumped out of the pool to greet her.

"Hi there! What the heck are you two up too? The whole house absolutely reeks of garlic!"

"Oh, it's the wild leeks, Mom. I found a huge patch down in the ravine and we've made creamy leek and potato soup."

"Supper's ready! Soup's on the table."

The kids tossed on their t-shirts and rushed to the table. Dad said grace and wished us *bon appétit!*

After a taste, Mom and my siblings made faces as they tried to retain their composure. Dad and I, well, we gobbled down the soup to preserve our dignity, but we might as well have been chomping on raw garlic cloves. Dad looked at me.

"André, this is *gak*. Those leeks of yours are way stronger than garlic. You don't have to eat the soup kids. André must have made a mistake again." He was surely referring to the time I made soup with wild *Russula* mushrooms, only to admit to not having identified the fungi correctly *after* everyone had eaten it.

Mom was opening all the windows and doors of the house, trying to ventilate my mistake. Thank God for Dad's wonderful *Supreme de Volaille Cordon Bleu* that followed, letting us forget the soup somewhat. And the homemade pumpkin pie.

Right after supper, I immediately jumped on my bike and pedalled off to the local library to search for the wild onions, plant in hand, to the disgusted looks of passersby who suffered an unwelcomed whiff. The error was soon revealed. I was in the possession of Crow garlic, *Allium vineale*, not wild leek, *Allium tricoccum*. Same genus, different species. Very different species. I rode home and walked back into the house, head low in shame as I admitted my mistake to my best friend in the whole world, dear old Dad. He sympathized.

"Chalk one up for experience!"

Relieved, I headed outside where the basket full of wild garlic lay. Not wanting to waste those plants, I decided to dry them to use sparingly on my next camping trips. With a needle and strong thread, I strung the plants into a five-metre necklace and hung them on the fence to dry. The next day, since the weatherman was predicting showers, I unlocked the garage and hung the semi-dry plants between the rafters.

Four days later, I was totally absorbed by Sir John Franklin's masterpiece on Arctic exploration when a loud shriek startled me. It was Mom.

"André-François Bourbeau! You come over here, and make it real quick too!"

Poor red-faced Gertrude was hovering just outside the garage door. As I approached I hit the garlicky wall. Double whammy!

That was but the tip of the wild-food-foraging-iceberg that served as a beacon, leading me into the icy but wonderful world of wilderness survival. I was haunted by a compulsion to experiment with all edible wild plants. One day it was the stench of skunk cabbage which fascinated me, the next it was the smashing of a burlap bag filled with hazelnuts onto the sidewalk to rid them of their prickly green covering, and the following it was the mysterious disappearance of the shaggy manes I had put in the refrigerator, finally explained by the fact that mushrooms in the *Coprinus* genus turn into ink overnight. Those were happy days, sipping staghorn sumac lemonade, savoring deep fried slices of battered giant puffball, and delighting in burdock stalk candy for dessert. I sure enjoyed this trip from Indian cucumber to hawthorn jelly — hiking the trail that would lead me to become the survival engineer I wished to be.

During my Bachelor of Arts and Science degree I enjoyed camping immensely and also pursued an interest in survival techniques by reading popular books on the subject. Then I enrolled in teacher training and completed a Bachelor of Education degree at the University of Toronto. I hoped for a job in physical education, my major, but ended up teaching math, the minor I had chosen because of the job opportunities.

"Anyone here like spicy Mexican food? Any tough guys around?"

I was facing my group of burly teenagers. Those youngsters volunteered to be part of our progressive high school's outdoor survival club. They responded to the posters I had plastered on the walls, à la Shackleton: "Join the outing club. Wild survival trips. Adventure guaranteed. Chance of no return!"

"Me! I *love* Mexican food!" replied the one square-shouldered heckler with bushy sideburns, my presumed target.

"Are you sure? This is really spicy, only for tough guys, for really manly men!"

"No problem, give me a huge bite!" said the disbeliever, trying to impress the gallery.

"Nah, just try this little wee piece first, then I'll give you more...."

With my razor sharp pocket knife, I sliced a paper-thin piece of the wild Jack-in-the-Pulpit root I had just unearthed and handed it to him. He didn't suspect that the root contained calcium oxalate crystals, called raphides, that are in fact microscopic needles that penetrate the skin of the tongue and lips. Hurts like hell. It takes fifteen seconds or so before you sense the pain.

"I don't feel a thing!" boasted the broad shouldered football player. "Wait, now I'm starting to feel it.... Holy shit! Water!" Water didn't help. He was maimed for at least half an hour.

"Told you it was spicy!"

"Yeah, but this is a hundred times worse than anything I've ever tasted!"

"Told you it was spicy! "

Needless to say, I had great listeners for the rest of my edible plants wilderness walk. Except the one sheepish statue with the watery eyes, standing immobile at the back....

Spicy

Nasty teacher! Shame on me. I remember oh-so-clearly the first time I myself felt the sting, both to body and ego, of Jack-in-the-Pulpit root. When I initially encountered the plant in the wilderness, I had but a vague recollection of having read about Indian turnip, the other common name for *Arisaema triphyllum*. So I pulled up the root, which looked delicious to me, having the texture of a potato. After peeling it, I stuffed a substantial piece of the root into my mouth and started chewing with hopeful anticipation. In the span of a few short seconds my taste buds had detected the short-lived pleasant sweetness, immediately followed by the acrid pain. I was convinced I was going to *die*! My tongue swelled up and the burning sensation was driving me crazy, especially since I had no clue what was happening to me. And I was alone.

After surviving this incident, I hit my books in short order and found out that good old Euell Gibbons, author of *Stalking the Wild Asparagus*, had suffered a similar fate. Gibbons went on to specify that once perfectly dry, the calcium oxalate crystals break down and the plant becomes palatable. Following his advice, I tried drying Arisaema roots in the attic, for months. But there always seemed to be just enough leftover moisture in the Indian turnip chips for the poison to kick in. Ouch again and again, at every optimistic taste.

These spicy failures encouraged me to pursue with further passion my study of edible wild plants. At one point, I had delved into the subject with such intensity that I could rattle off by heart the scientific name of just about every Ontario plant, without having made the slightest effort at memorization. The awe-inspiring variety of flavours cannot leave one indifferent. Wild ginger! Wild carrot! May apples! Violet leaf salad! These finds quickly erased the turnip incident. That's when I truly fell in love with nature and its intricacies.

Five

The Swamp Tramp

"Where there are no swamps there are no frogs."
— German proverb

After that first edible plant field trip with my brave outing club students, they were convinced my posters were true — there really was a chance of non-return. News of how I had checkmated "Big Dave" had spread through the school like wildfire. Eighty-five teenagers showed up at the second meeting. Too many. An on-the-spot decision to charge twenty bucks to become official members thinned the crowd to twenty-two and the club was bestowed its first $440 worth of camping equipment. A few pack-a-thons, egg-a-thons, and outdoor auctions later, we had gathered enough stuff to head out on our first multi-day survival camping trip, with me as the twenty-two year old "leader." A chief full of ambition and enthusiasm, yes, but with no leadership skills whatsoever.

By that point in my life at least I was already technically accomplished, or so I thought. After all, it had been five full years since my first winter camping trip. Jules Verne had not only enticed me to learn edible plants and fire starting skills, but also to make rope from basswood bark, tan a skunk skin, weave cattail hats, make soap from fat and ashes, build deadfall traps, and so on. My learning strategy was always the same: trial and error. The hard way. Paint hot pine tar onto the wooden skis and then ski without gliding all winter until it wears off. Or go to the end of the road during spring thaw and get my old 1968 Rambler hopelessly stuck in deep mud. How else would I have discovered the jack-the-car-and-shove-it-out-of-the-rut technique?

What better way to learn than to put myself in a predicament, I thought. Go camping and "accidentally" forget the pot. Or the matches. Or the tent poles. Or the food. By gradually removing gear from the equipment list, I had to find substitutes. If I go far enough into the wilderness without rope, ties, wire, straps, or belts, I reasoned, well then I'll become proficient at the inverted double twist rope-making technique. And if I go out with raw hamburger and no matches nor knife, I will by necessity have to start my friction fire like a caveman. To be honest, this tactic worked — to some extent. I did eat a lot of raw hamburger, and did sleep in many half-erected and soaking tents. But I also developed solid improvising skills. I became handy, deft, and adroit — at the cost of being scratched, maimed, or beaten down by illness.

But, as I was about to find out, outdoor technique has little to do with leadership. I had everything to learn about managing kids, especially to their parents' standard.

Jacking the car and shoving it to get the wheels out of a rut.

"Who wants to go wilderness camping this weekend?"

"Me!"

"Me!"

"Me too!"

Sixteen hands were waving near the ceiling. The jaws of the other six teenagers in the room drooped as low as their dangling arms; their parents had planned an altogether different kind of trip, mostly to visit boring relatives.

"Okay, let's start planning. We'll go to a place I know called Kimball Lake; it's near Algonquin Park. For transport, we can use my van. Can all you guys and gals bum food from your parents? Can you beg, borrow, or steal a sleeping bag and a foam mat? We'll use the club pots and tents. I'll take care of the dishes, utensils and cutlery."

In no time, all was settled. Between club gear and personal gear, the essentials were accounted for. On this first trip, my secret plan was to "accidentally" forget the eating gear, which I did not possess. I figured the kids would have fun making wooden cutlery and birchbark dishes.

"We'll leave from here this Friday at 4:03 p.m. Be ready! Tell your parents we'll return Sunday night after supper."

"We'll be here!" The excitement was so tangible you could cut it with a knife into sixteen slices. As they left, I too was delighted, but as I headed home my enthusiasm wilted with the recollection of a math exam I had to prepare for the morrow.

A few X's on the calendar, and the Friday afternoon gathering around the pile of packsacks had arrived.

"It's 3:45. Eighteen minutes left to pack up the van before we leave at 4:03! Let's get a move on!"

"Yes sir!"

I owned a former Bell Canada van, completely empty except for the two front seats. We shoved and squeezed all the packsacks in side by side on the floor of the van, having first removed the hanging blue Ensolite foam pad rolls. First layer. We unrolled all the foam pads and crisscrossed them over the packsacks. Second layer. Then all the teenagers piled in, third layer. We took off to the blaring sound of Creedence Clearwater Revival's *Born on the Bayou*.

"It's 4:03! We did it! Let's hear a big cheer!"

"Yahoo! Born on the bayou!"

"*They're off!* yelled the monkey as he backed into the lawnmower!"

"I don't get it."

"They're off. His gonads. Get it now?"

Spirits soared high as we took the ramp onto Highway 400 north to where the pavement turns to sand, as Neil Young's song so eloquently states. Three and a half hours later, we were at the end of Bear Lake Road as the sleepy crowd jumped out, running to hide behind convenient trees for relief.

Our plan was to hike around Kimball Lake in a counterclockwise direction, with that night's camp at the nearby western creek and the next day's destination on the point of land to the east on the other side of the lake. From there on Sunday we would head north, cross the creek at the northeast corner, and hit the bush trail that runs parallel to the north shore back to Bear Lake Road.

We hiked the few hundred metres to the first camp spot without incident and able bodies set up the tents while others tended the fire and pulled out the supper. Everyone was starving; it was almost 9:30 p.m. and the sun had already dipped below the horizon.

"Hey old man, where are the dishes and cutlery?" I loved it when the students called me that. At twenty-two years old, it was quite the compliment. It made me believe I owned wisdom.

"In the green pack over there." My magician training made me a good liar.

"Can't find them!"

"What? Let me look. Darn, must have left them beside the door when I left home."

"What are we gonna do?"

"We'll just have to eat the rice on a piece of bark and whittle our own cutlery. I'll show you how to make automatic chopsticks. By bending a maple twig in half until it breaks but stays attached, and cutting the ends even to the correct length, the legs of the chopsticks keep springing apart when you squeeze the two ends together. See? Automatic!"

As the rice cooked on the cheerful fire, everyone whittled his or her own eating gear. It's magic.

"All you primitive Chinese guys look terrific!"

The next day, we were up bright and early in spite of the late tuck-in. We got to camp number two on the point just before lunch. I suddenly realized that the big kids needed a challenge to keep them busy.

"Hey, let's make rafts! Then tomorrow we can cross the lake and we won't have to hike all the way around."

The packs were heavy and the suggestion was met with enthusiasm.

"How do we make one?"

"I don't know. Let's experiment!" I'd never made a raft before. This was great. I could boast sixteen research assistants helping me.

Thusly, after camp was set up and lunch gobbled down, the rest of the day was spent tossing wood chips with our axes and rounding up massive logs near the shore. Each group of four scrounged up every piece of rope from their gear. As darkness fell on Kimball Lake's point, no raft was yet worthy of its name.

"Boy, this is going to be harder than it looks!"

"Yep!"

The entire morning of the next day was spent desperately trying to get the logs of the rafts tied together. We discovered that semi-dry logs aren't very buoyant. Nor are hardwoods. The best raft floated about the depth of boots underwater when loaded with four teenagers and gear. After lunch, more dry logs were gathered to fix the problem. Time flew until my watch read 3:15 p.m. Time to go. It was a long way around the lake back to the van.

"We'll have to pack it in gang! It's getting late and we have a two hour hike and a long drive ahead of us!"

"But Sir! Our rafts are almost finished!" They were all yelling and complaining in unison.

"Okay, I'll give you another half hour, but if the rafts aren't floating by then we'll have to walk."

The half hour stretched to forty-five minutes, then to an hour. All rafts were an utter failure and the reality check helped me convince the youngsters to abandon their project. I was actually relieved, as I hadn't even considered the fact that we were without life jackets. It took quite a while to recuperate the ropes and clean up the site. Finally we were on our way, but it was already suppertime. We munched down on the "all-dats" leftover, tossed on our packs, adjusted our tumplines, and headed north toward the trail. In no time we got to the creek. Surprise! There was no creek. The beavers had turned it into a major swamp. It was just past seven.

"Hey gang, I wasn't expecting this creek to be any different than the one we vaulted over easily at the other end of the lake. But it'll take too long to go back around. Let's hike east along the swamp to see if we can find a crossing point."

After half an hour of tough hiking amidst thick bushes, there was still no sign of a beaver dam. The swamp seemed to go on forever. We rambled on to the east. This was tough going, some of the kids were exhibiting pre-whimper non-verbal signs. I was really at a loss. What if this swamp continued for miles? Further on I noticed the swamp narrow, but there was still no end in sight. I had to make a decision. I put on my magician's smile and blurted out with optimism.

"Hey you guys! Remember the posters? Welcome to the swamp tramp! Guaranteed adventure! This is where we cross."

Without waiting for a response, I was up to my knees in oozing mud, heading north. I figured we didn't have a choice; the sun was four fingers above the horizon, which meant about an hour of real light left. And as usual, I had voluntarily "forgotten" the flashlights. The incredulous youngsters were following me. Good.

As the water reached my waist, all was well, we were halfway across. Then we hit the channel, where the water was deeper still. I was up to my armpits. This won't do! Wait, it's less deep now, I think we'll make it. Then a tremendous shout split the air.

"Sir! Help! My running shoe got stuck in the mud and I can't find it!"

That short kid had water up to his neck.

"Forget the shoe! We'll make you a new one when we get to the other side. We're almost there! Just keep going! Let's go gang, you're doing great!"

Once on shore, I'd never been so happy to pick cattail, bur reed, and water lily remnants from my clothes. Reminded me of the last time I went hunting for bullfrogs, minus the hordes of mosquitoes. At least there were no bugs in late September.

The Swamp Tramp

"Okay everybody, wring out your clothes and put them back on."

Nobody had extra clothing; they all listened too carefully when I suggested that tough survival guys don't bring any. Cold was not a factor, though; it was fairly warm out. I headed over to the short one-shoed kid. I instructed him to go barefoot inside his shoe and give me his sock. I stuffed birchbark in the sock and told him to put it on over his other sock so that the bark acted as a sole. Then I wrapped a rope around and around the "shoe" to hold it together and prevent wear and tear of the sock. I had practised this on prior survival experiments when I was pretending that my boots had been burned in the fire.

"Let's head north to the trail!"

Miniature compass in hand, I vaguely followed the needle. I figured it couldn't be much more than half a kilometre to the trail. Then it would be smooth sailing to the truck. So off I lead into the wild, like Abraham in the desert followed by his caravan of sixteen acne-prone camels. My reverie was interrupted as I hit the evergreen wall. The forest there was so thick we couldn't walk through. So we crawled. And it was getting dark. The balsam fir needles and twigs stuck to our wet clothing like glue. After a while of this commando-style escapade we were back into maple forest. But darkness has fully settled in and I could only read my compass by flicking my disposable lighter every few steps. On one such flick, I realized to my horror that my next step would have been into the void, over the edge of a short but deadly rock cliff. I didn't say a word as I veered off to the west, flicking my way down and guiding my group out of the danger zone.

The blind leading the blind. After what seemed an eternity to a bunch of dazed, dirty, and wet swashbucklers, who were no doubt under the impression that their ocean of troubles would never end, the pirate ship finally landed on the trail, to the incredible cheers of all present.

The next hour's dancing walk to the truck was the most incredible display of joy and camaraderie I have ever witnessed. Those teens were beaming! On cue, the moon rose to celebrate our victory and light our way as we merrily sung and chanted. It was past eleven o'clock when the van was finally loaded.

On the way back to Toronto, near Orillia, we were intercepted by the flashing sirens of the Ontario Provincial Police. Was I going too fast? The policeman approached my door. All of the teens were dead asleep at the back.

"Are you Mr. Warble?"

"No sir! Definitely not."

"Okay. Sorry to trouble you."

I drove into the schoolyard at 2:30 a.m. I couldn't fathom disturbing all the parents at that time of night so I decided to drive all the kids to their homes. As I delivered the last one to his trembling mom at 3:30 a.m. in his underwear, overloaded by all his pell-mell stuff including his wet and filthy jeans which repeatedly fell on the driveway, it finally dawned on me that maybe I wasn't so "cool" from the parents point of view. Especially after the very dirty look I just got from that mom!

The next morning, right in the middle of my first math class, the announcement speaker came on calling my name and requesting my immediate presence at the principal's office. This cued the *oohs* and *aahs* from the students and reminded me of Mom when she called out my full name. I walked into the principal's office with a wide grin. It's obvious to me now that I'm a dad, but it wasn't then — naturally all of the parents had been phoning the principal at his home as of ten o'clock the night before. And the cops had been looking all over for us; in fact, Mr. Warble had been the officer's phonetic rendition of Bourbeau. He had been expecting a van full of hectic kids, I suppose.

"Why didn't you call as soon as you got out of the bush?"

"I didn't have a quarter for the phone booth and I didn't want to disturb anyone at such a late hour."

"You can imagine that I had a hard time convincing the parents not to worry!"

"Real sorry about that. I learned my lesson. Won't happen again."

"You know we'll have to bring this up at the next board meeting."

What was I thinking!? It was as if my survival training and my wish to be independent from society had erased all administrative concerns from my mind. Now I was in real trouble, and my wished-for career as an outdoor educator was compromised. The principal was fond of me, I knew, ever since I shook his hand like a pair of vice-grips when he interviewed me for the job; he knew me as a man of character. He was fond of me, but was it enough to get me out of this mess?

In between classes, a couple of my participants came to enquire why the principal had called upon me. I suppose they knew, having heard their parents' comments during breakfast. I confided in them. That afternoon after school, unbeknownst to me, the entire club had asked for an interview with the principal. They apparently praised my leadership skills and stressed the fact that at no moment were they ever even slightly worried about their safety.

That's what saved me from utter humiliation — that time.

Six

Up Here, Sir!

"There's none so small but you his aid may need."
— Jean de La Fontaine

After my first real-life lesson in confronting what Norman Mailer aptly named "those moral wildernesses of civilized life," I calmed down a bit. But not for long. My teenage buddies became inspiring partners in adventure. Together we participated in ever more eccentric trips. Sometimes we would impatiently wait for the school bell to ring on a Wednesday afternoon so we could hop in the van to go survival camping only to return in time for Thursday's morning class, dressed in suit and tie and having met the objective of remaining spotlessly clean. Or we would head out weekends to experiment, either by leaving part of our equipment behind, or by inventing scenarios which would simulate accidents, such as flipping over while canoeing and reaching shore soaking wet.

Our outing club was so successful that I convinced the principal and board of directors that Thornlea High School should become the first to develop an outdoor education curriculum. Thusly, for my second year as teacher, I ended up professing mathematics in the morning and conducting outdoor activities every afternoon. Plus survival weekends. On one particular trip that fall, my intention was to test a technique unearthed in one of the ancient woodsmanship books I had just gobbled up.

"Hey gang! This old book mentions that we can make a fire last longer by setting up a system that lets logs roll down an incline into the fire automatically, triggered by the burning of the previous log. That way we can sleep by the fire the whole night without waking up. Want to go try it this weekend?"

Just before lunch Saturday we parked the van at the end of Bear Lake Road as usual, and headed to the shores of Livingstone Lake. It was early November, quite cold, and the wind was shaking the trees like a rag doll in a dog's jaw. So we portaged some "hypocritical sleeping bags" in a couple of army sacks, just in case. These were sealed with tie wraps. I was so stubborn I wouldn't have opened them even if I were dying. But these kids have parents.

After cooking up some bannock by carefully wrapping spirals of dough on sticks and stuffing our faces with peanut butter and jam, we tossed the axe sheaths aside and got to work. I wanted to sleep without waking up each hour to toss logs on the fire, and I'd managed to convince my buddies that it was possible. We spent the afternoon chopping huge maple logs in two-metre lengths and lugging them to our chosen spot at the base of a hill, which would serve as incline for the rolling logs. It was backbreaking work, so much so that we were all down to t-shirts and still sweating, despite the frosty air. By suppertime, which at that time of year coincides with darkness, we were all collapsing, testing our thick fir bough beds in front of roaring fires. The wind was howling and screaming like a pack of wild wolves, and the turbulent smoke became their shadowy ghosts, which made our lives miserable.

Finally satisfied with my automatic-log-rolling-into-the-fire-as-the-previous-one-burns setup, albeit with serious doubts as to its intended midnight functionality, I decided to go check the progress of the teams installed nearby. All presented a glow of satisfaction as I scrutinized and approved the results of their hard labour — cords of firewood neatly stacked in preparation for a night without sleeping gear. I also marvelled at their ingenuity in inventing systems to meet my challenge to add logs to the fire with the simple pull of a string while reclined in their beds.

Soon they were busy getting supper ready, levelling the bright ruby coals where they would directly toss their "caveman steaks." Gourmets, beware. Nothing in the world tastes better than a caveman steak! As the meat hit those coals, it smothered them and prevented scorching, while the smoky heat below imparted an incredible flavour. I couldn't wait to go cook mine. But just then I realized that I hadn't seen one of the youngsters in a while.

"Where's Blake? Anyone seen him?"

"Nope, not me."

"Me neither."

"Nor me."

"Who was he teamed up with?"

"Nobody. He was the odd man out, remember? He must be alone."

Not proud of myself. Why didn't I check up on Blake? Blake! Why him? I would never say it out loud, but that pesky kid drove me nuts. He was, well, skinny, nerdy, and lazier than a sleepy sloth. But brilliant, I must admit, always asking me to justify what I uttered. I yelled out his name to no avail. Why couldn't he just follow directions like the other teens? Where in the world was he?

As not to alarm the group I conducted the primary search myself, probing along the paths of least resistance, calling out Blake's name. Half an hour was spent looking here and there, getting more worried by the minute. I searched the entire valley; no sign of him. The fierce wind drowning out my voice sure didn't help.

I climbed halfway up the hill and shouted some more. Why in the world would anyone, even Blake, want to set up camp up where the wind blows fiercer still? But since he wasn't down in the valley, he must be up here somewhere! I walked along the contour line, trying to avoid hiking up and down to save my energy. I'd come full circle around the valley, almost back to camp. I yelled once more.

"Blake! Blake!"

"Up here, sir!" came a weak voice through the rustling leaves.

Most relieved, as it was now dark, I made my way up toward the youngster, hoping he was okay. On the flat at the top of the hill, the forest changed drastically. I progressed through a stand of small pine trees, obviously a plantation following some logging company's harvest. Guided by the feeble light from his campfire, I finally caught up to the elusive Blake.

"What the heck are you doing up here? There's no firewood in this plantation! You'll freeze to death!"

"Nah, there's lots of these little dry branches. See, I just snap them off the trees. There are tons of them."

I couldn't believe what I was seeing. There sat Blake on a stack of dry branches, back leaning against an arm-sized evergreen, feeding tiny sticks to his tiny fire. On the end of a green sapling, he was calmly roasting a marshmallow-sized piece of steak, shish-kabob style. To tell the truth, he seemed just fine! And to my surprise, there wasn't the slightest bit of wind in the dense plantation. We chatted for a while, until my steak called out for me.

"You want to come down and share my huge hot fire?"

"Nope, I'll be fine up here, sir!"

"Suit yourself, if you want to suffer. If you change your mind during the night, you know where to find me."

"Okay. See you in the morning."

I scrambled back down to my own camp, no longer worried, since I could pinpoint Blake's location only a couple of hundred metres uphill. My hunk of beef sure hit the spot, and after another round to visit everyone, I crashed onto my personal pile of boughs by the fire, exhausted. Just as I closed my eyes, I heard a log fall into the fire to replace the one that had just burned through, and a satisfied smile slid across my face.

The pleasure was short-lived. Before midnight, I was up and nosing about, attempting to figure out why the next log wasn't as cooperative. I tossed the partly burned log leftovers onto the remaining coals, fanning with my felt hat to coax them back into a bright fire. The blustery weather had not let up, and despite the warmth radiating in front of me, the wind continued its chilly attack from behind. I retreated back to my horizontal position, tossing and turning without finding any real comfort.

Half an hour later, shivering slightly, I decided I might as well check up on the four other teams. All of them were suffering the same cruel fate. Incredibly huge piles of wood had already been consumed, and it was obvious they would never have enough to get through the night. I suggested they group around two fires only. The three-quarter moon shined on the proceedings as they transferred firewood and bough beds to the new locations.

I headed up the hill to check up on Blake. Locating his camp was easy, having memorized the prominent marks along the way: past the forked tree, around the huge boulder, between the twin stumps, and then under the arched tree to the pine plantation on the crest. I approached slowly, curious to see what he was up to. He was up to nothing at all. In fact, he was just lying there sleeping! He was wearing that stupid thick jacket of his, fully patched with car racing decals; he wouldn't listen during the dress-in-onion-layers class. But there he was, mouth agape, sprawled over the thick pine-needle covered ground, his miniature fire dead out. I sat down, flabbergasted, mesmerized by the moon's eerie shadows, the shrieking of the wind above, and the calmness there in Blake's protected haven.

Still pensive, I stumbled slowly back to my own huge campfire where the wind was creating havoc. So much work for so little comfort! I sucked it up and endured my fate until first morning light. By then, all eight of my partners were sitting by the same fire, eating toast. The wind had died. Soon, Blake joined us. He told of how he got slightly cold during the night and had to get up to make another fire. No one listened to the odd man out. And I, much too proud to admit the error of my ways, congratulated everyone on their hard work. But deep inside, past my blurry and smoke-sore eyes, I was well aware of having just learned the most important of wilderness secrets — avoid wind at all costs — and dumbfounded that Mother Nature would send a fourteen-year-old messiah named Blake to teach it to me.

It took until lunch to burn all of the remaining wood to ashes, drown them with creek water, and disperse them by shovel over a wide area. Finally, we buried the site with forest litter since I make it a matter of pride to always leave camp so no one knows we were there.

Little wonder, all the youngsters were sound asleep in the back of the van as we drove home. All except pesky Blake who sat in front with me, dressed in his stupid racing coat, bugging me as usual.

My adventures as a high school teacher ended abruptly one day during math class when some kid asked me the use of the sine law I was teaching. I hadn't a clue! This disturbed me greatly and that night, as I reflected upon my life, I realized that I had spent just about all of it within four blank walls — first as a student, now as a teacher. Just having turned twenty-four, I couldn't imagine there wasn't more to life than that. So I planted a "For Sale" sign in front of the country house I had bought and renovated in the village of Whitevale and announced that I would quit my job at the end of the year, to the chagrin of my student friends.

To celebrate my departure, I decided to win the annual teacher tricycle race. The year before I had been beaten by Miss Pickering, the ten year champ. Unbeknownst to me, she had enlisted a pack of football ruffians who held everyone else back. But now I could convince my own team of outdoor tough guys to restrain her too.

I decided to psych her out. I made myself a souped-up tricycle, with extra long handlebars and seat post, all decorated. Then I scrounged up a leather pilot's helmet, Snoopy style, and a long scarf. Every day for a month before the race, I would screech down the school hallways in my super-trike, making sure to spin the front wheel while passing by her class.

The week before the big event, as I chalked up my daily ride, some student in the crowd blurted out: "There goes Evel André Knievel!" So I replied without thinking: "Yeah and Friday I will jump over forty glasses of water!" What I hadn't realized was that even with forty small paper cups, the distance to jump was well over two metres. On a tricycle!

Back home that evening I built a plywood ramp, and after a few practice jumps over half that distance I was already most concerned for my family jewels. But by then the students had put signs up all over the school walls and had even invited the media. On Friday the whole school turned up for the anticipated jump, the principal having suspended the last class for the occasion. To make a long story short, I was absolutely petrified. After the *oohs* and *aahs* of my last-second swoops to avoid the ramp, I no longer had the choice but to muster up the courage and attempt the jump. I made it — barely. One back wheel hit the edge of the landing ramp and sent me swerving left then right, crashing into the crowd. By a stroke of pure luck, no one was hurt, not even me. The cheers blew me away as I was carried down the hall by the students as if a national hero.

The next day, the newspaper headlines read: "Is it a bird, is it a plane? No, it's a flying teacher!" And yes, I did rob Miss Pickering's crown the following week, if only by a nose.

Today, as I look with hindsight at young André, I can't help but smile. Passionate yes, fanatical even, but performance-driven by an ego the size of a Goodyear Blimp. This fat characteristic followed me like a pest into adult life. When my friends tease me about this, I jokingly reply that I cannot be humble, for then I would be perfect!

Seven

A Sheep

> "People, like sheep, tend to follow a leader — occasionally in the right direction."
> — Alexander Chase

The rest of my school year at Thornlea Secondary School was rather uneventful. Except for one day in late March, when I was approached at lunch by a fellow teacher and avid outdoorsman, Frank, with whom I had shared a couple of wilderness canoe trips.

"Hey André, what are you doing after school today?"

"Nothing special, Frank. It's Monday."

"Do you think you could come with me to visit a piece of land I'm thinking of buying? It's about forty minutes northeast of here. We should be back by suppertime. I'd like your advice about the quality of the ecosystem there."

"Sure, no problem! See you at 3:30."

As soon as I answered some zealous question posed after the end-of-day bell, I dashed out to the parking lot where Frank was waiting in his fiery red pickup.

"Just a sec, Frank. I'll just grab the rubber galoshes I keep in the van. There might still be a bit of snow on the ground."

We headed east from Thornhill, the spring sun blazing through the rear windshield, and after half an hour or so we veered straight north. Soon we hit a country road, then another, and after a few corners and twists and turns, I spotted the realty sign. As we headed out into the woods to check the place out, something felt strange. Then I got it. We were dressed in suits and ties, as per school policy.

Beautiful piece of land. And expansive. The place had nothing in common with the city we just came from. Huge trees, rolling hills; it was a real pleasure to be there, as the sun played hide and seek behind the clouds. I was not aware there were such parcels for sale so near Toronto.

"Frank, this place must cost a fortune!"

"Well it's not that bad actually. You see, the land for sale isn't that big, but it borders on the Glen Major Conservation Area we're in now. That's what makes it such a good deal and why I'm so excited. What about the environment around here? Anything special?"

Frank pointed out different species of deciduous trees such as oak, ironwood, beech, or sugar maple, and I confirmed his identification and their approximate age of sixty years. As we climbed from a slight valley, a willow attracted my attention. They are rarely found at the top of a hill, since they prefer moist ground; this one must have had long roots. As we continued our exploration, Frank asked me to help him identify the animal tracks we encountered and some wilted plants he wasn't familiar with. He was quite curious, and as we sat on an overturned hemlock to take a break, he gobbled up my improvised lecture on the ecosystem around us, all the while taking pictures.

I followed Frank's heels down into the next dip and up the other side, and noticed the well-hidden sun was winning its game of hide and seek. We'd hiked quite a ways through some wonderfully fine woodland. We crossed another insignificant valley, then climbed up a short hill. That's when I spotted the willow. Two willows on top of a hill on the same day? Hmm.

"Say Frank, do you know where we are?"

"Well, actually, err, not exactly."

"Why didn't you tell me? I've been following you like a sheep, and we've come full circle! We crossed this willow about 45 minutes ago!"

A Sheep

"Are you sure? I thought we were heading back to the truck!"

"Well obviously neither one of us knows where the truck is, and I'm getting hungry, as I'm sure you are. This conservation area can't be that big. Let's at least just walk in a straight line, and we'll eventually come to a road."

Lost in the woods. For real. I loved it! Then I realized I didn't have a way to light a fire on me. Ooh, that made me *so* mad. How could I, of all people, supposed survival expert, be caught in the woods without a means to start fire? Then I recalled that that morning I was heading to a city job in suit and tie. Still, I'll never again be caught without a butane lighter in my pockets. Never! Never ever! I swear!

"Got any matches Frank?"

"Nope, don't smoke."

"Me neither."

I did have my miniature pocket knife though, and a shoelace, so I briefly considered starting a fire by rubbing sticks. That would take a couple of hours at least, and it would be dark by then. Not a good option. To make matters worse, the clouds were darkening and the barometer was obviously falling, as was the temperature. It would rain that night, no doubt.

No panic. We picked a direction at random and chose three trees that lined up like telephone poles. I stood at the second tree while Frank hiked forward to the third. From there I lined up the second and third tree to spot a fourth one, and guided Frank toward to it. Then we repeated the process, over and over again. That way we were walking in a straight line that would eventually get us out of trouble.

After half an hour of the post-to-post proceedings, I felt uneasy. We'd been walking for two kilometres and there was no sign of civilization yet. Undoubtedly, we had chosen a wrong direction. And as darkness approached it was starting to drizzle. We had no choice but to continue; it was our only chance of drinking tea that night, instead of crouching shivering under some hastily made shelter until morning. Yet as I stumbled along, I scrutinized my surroundings for potential camping spots. Lots of firewood here. Darn! No possibility of fire!

I was getting cold. Without the slightest hesitation, I ripped the sleeves off my suit jacket and used my tie to knot them over my head as a hat. I'd done that before. I don't care, I always buy my wool Harris Tweed jackets at second hand stores for five bucks, and nobody ever notices the difference. Frank thought I was losing it. I reassured him, explaining the importance of a hat for retaining body heat. As I passed by a fir tree, I grabbed some of the lower branches that had remained dry and stuffed them inside my jacket for added insulation. Frank was scratching his head. He wasn't sure whether or not he should imitate me.

He didn't have to. Just four tree-line-ups later, we emerged onto the anticipated road, much to our delight. Our worries weren't over, for we didn't have a clue which way we would stumble upon the truck. And there was not a tire mark to be seen. We tossed a coin and went left, finally finding the truck, after many back-and-forths, near midnight. We were starving, wet, cold, and tired. But the next morning found us preaching in front of our respective classes, as if nothing had happened.

I was playing hockey in a local arena. Like any Québécois worthy of his or her name, I love trying to out-manoeuvre my garage league's goalie. In the dressing room, my fellow players still tease me every time they notice the disposable lighter taped to the belt of my hockey pants. Proof I've been respecting my vow. There's nothing like a close call to drive home Baden Powell's message to the Scouting movement: "Be prepared!"

My stint as a high school teacher had come to an end. I spent the following summer in Algonquin Park, as I had done for the last five years. Dad had started up a family business managing kitchens for many Ontario summer camps, with the boring name of GB Catering Service, based on his and mom's initials. GB's philosophy was to promote camping by providing high-class homemade meals, including daily fresh baked bread and pastries. As eldest son, I held one of the key positions in the company.

My summer jobs had always evolved in kitchens. Starting at the age of fourteen, I had climbed the ladder from pot washer to chef. I served my apprenticeship in one of Toronto airport's super kitchens. I'll always remember the first day at work, when the executive chef looked me up and down: "See those eight hundred chickens I just cooked? Bone them

A Sheep

Survival dressed in suit and tie.

out!" The next day I broke forty cases of thirty dozen eggs each into huge omelet mixers, and the day after that I grilled sausages — for twelve hours straight.

At twenty years old, when Dad coerced me into being chef at the famous Taylor Statten Camps in Algonquin Park, during its glory days when the dining hall was jam packed with close to five hundred people, at least I was used to bulk cooking. And I was tickled to be in mosquito country, for that meant endless possibilities of practising canoeing, archery, and bushcraft — with the added bonus of enjoying Dr. Tay himself as mentor. Too bad there was so little time left after those gruelling long days slaving over the stove. Secretly, over the next five summers, although I had been promoted from chef to kitchen administrator, I wished I could switch places with one of those fortunate camp counsellors who went off on month-long canoe trips to Quetico or to Kipawa parks. But the expertise I had gained as the lone soul in charge of computerized food planning for the hundreds of canoe trips under GB's responsibility kept me at Dad's side.

In early fall, having finally sold my house, including the antique furnishings, I gave all the possessions that wouldn't fit into my van to friends and family and headed off to the University of Northern Colorado, where I had been accepted into a unique Master's program in outdoor education. I was enthralled to say the least, so very tickled to have found a curriculum designed specifically for me, perfectly convinced I would learn a great deal more about wilderness survival and open-air life.

Eight

The Farmer

"Adversity is like a strong wind. It tears away from us all but the things that cannot be torn, so that we see ourselves as we really are."

— Arthur Golden

The van was jam packed as I sped along Highway 80 through Iowa and then Nebraska, on my way to Colorado to check out the worth of those impossibly steep tuition fees I had just dished out as a foreign student. My reverie of seeing the Rocky Mountains for the first time was suddenly interrupted by engine noise, which made me stop at a nearby garage. No way I was going to spend three hundred and fifty dollars to change a burned out valve on this old van! "How much just to grind the valve?" I asked. Five bucks sounded a lot better to me, so I found a vacant lot nearby and got to work taking that engine apart. Six hours later, I was in the possession of a freshly ground valve but the necessary new head gasket was priced at over sixty dollars. I did without. After spraying the old head gasket with a can of aluminum paint, I torqued the bolts and was off again, albeit with dirty hands. Twenty thousand kilometres later, I knew that paint had done the trick. The rest of the uneventful trip was spent reflecting on how any bit of practical knowledge can make a difference when facing a specific survival situation, be it to execute a mechanical repair, pick a lock, or even fly a plane. So much to learn!

Once at the University of Northern Colorado, among all of my classes I was really looking forward to the Outdoor Survival 508 course. I was dumbfounded when I found out who they had chosen as part-time lecturer — me. But I sure didn't lecture much. Preferring to experiment, I would take fellow students out into the prairie woodlots without gear. There they tried their hand at such skills as friction fires, tool-making by breaking stones, sucking water from wild asparagus shoots, and fishing for carp with homemade spears. On one occasion we made a huge grass debris hut so all twenty-four of us could sleep in it. Cuddled up like spoons, we affronted the cold night by turning over and rotating every half hour. Only the two people on the ends had to shiver for their own thirty-minute stint.

Other than this one memorable occasion, my first year as a graduate student had proven uneventful, saddled with adrenaline-devoid classes such as Philosophy of Outdoor Education and Administration of Outdoor Education Programs. At least I developed wonderful relationships with fellow students who complained as I did of boredom, and our

The twenty-four-person debris hut we slept in.

fun was found sharing rock climbing and mountain hiking adventures, which are Colorado's state pastimes.

As I was about to graduate with my Master's degree, I decided to imitate many of my buddies who were continuing at the doctoral level. It so happened that the University of Northern Colorado offered at the time a world-unique program called the School of Educational Change and Development, which permitted students to present a special doctoral project of their own creation. If I could convince seven professors to accept my project, it would become a signed deal. I proposed a plan that borrowed courses from a vast selection of the University's existing programs that I felt would help me pursue my quest for survival knowledge. For example, I would delve into North American Indian lore in the history department, learn to flake stone tools in the lithic technology courses of the archeology program, study plants in depth through systematic botany, and learn to identify and work with useful minerals with geology students at my side. My plan also included trips to several Third World countries to study local primitive survival techniques; I would go off to Africa, India, and Central America. Finally, my doctoral program would be completed by conducting week-long survival experiments without gear in each of the four seasons, to personally test the panoply of techniques I would expose in my thesis. After many discussions and adjustments, the professors finally accepted my project and I became the school's first doctoral student in the field of wilderness survival education. I felt honoured.

I kept journals of the four survival experiments I conducted. The first was in late October. The weatherman announced a warning of early snow, which would add to the adventure. Fine with me. I sure didn't suspect it would turn out to be a record-breaking storm, dumping half a metre of fresh powder within a couple of hours.

"Hey Bill, you can drop me off here on this bridge."

"Are you sure you're going to be okay for a whole week? This blowing snow looks pretty bad to me!"

"Don't worry, Bill, you know I've done a lot worse. Come and check up on me in three days. We'll be able to yell at each other from the other side of the river where you usually drop me off. You'll see, I'll be in great shape."

"You're crazier than I thought!"

I headed to the confluence of the South Platte and Cache la Poudre rivers, about four kilometres away across some farmer's fields. In no time I would be in the sheltered woods on the long point I'd camped at before. It would have been easier to get there by driving to the roundabout on the other side of the river and wading across, but the water was too cold at that time of year.

I headed east, using the snow blustering obliquely from my left to guide me. After two minutes hiking in ankle deep powder, the whiteout had erased the picture of the bridge behind me. I wasn't the least bit concerned about direction finding, since I was standing on a point of land between two rivers; as soon as I reached either I would follow its shore to the confluence. So I pushed on.

The wind somehow seemed sinister, sending swirls of the white stuff to caress me like so many wispy phantom fingers. On the ground, any irregularities caused the snow to pile up into drifting banks. Soon I reached the edge of the field, where a drainage ditch filled with water stopped me. I followed this depression, hoping to find an easy place to cross, but to no avail. I spotted a fence and manage to pry loose one of its posts, which I used as a bridge. As I hopped across, the semi-rotten grey wood broke in half and fell into the water, but at least I avoided a soaking. I realized that I'd "burnt my bridge behind me," finding the expression ironical in this freezing context.

In the short twenty minutes it took to fetch the fence post to cross the ditch, the snow had accumulated to just below my knees. The field I was crossing had recently been plowed, and the ruts rendered the walking treacherous. I slowed down, stressed by the thought of a sprained ankle. After an hour confronting the blizzard, I was no longer so sure of myself. Nature has a way of dealing with overconfidence. The drifts got up to a metre high, and my pace was ever-increasingly sluggish. Before I knew it, I'd slowed to a crawl, and I couldn't see ten paces in front of me. My pants were soaking wet up to the crotch.

For this week-long experiment, I was carrying a few items, since I was pretending to be a seventeenth-century Métis travelling between two villages. Thusly, my equipment list included a pound of wild rice and an equivalent amount of animal fat stuffed into a tin can, a belt knife, flint and steel with tinder in a waterproof tin box, a one metre by half metre piece of fur, and the clothes on my back: leather boots, one pair of socks, cotton pants, underwear, belt, shirt, wool sweater, leather jacket, wool mitts, and a toque. I wore eyeglasses, a compromise I have to accept, as well as a watch. My pot, food, and tinder box were wrapped in the sheepskin, tied into a tight bundle with two leather thongs. Other than that, the only items I carried were a notepad, pencil, and my wallet, which would be needed when I got out of there the following week.

As the storm raged on, I was wasting considerable energy plowing through the drifts and fighting the cold's attack, well aware that soon I would be in bad shape. I was no longer a happy camper, but I had to plug ahead. So I counted step sequences: one, one two, one two three, one two three four … Up to ten, then backwards to one. This helped.

After a few such sequences, the apparition of a huge willow tree became my first sign of progress and my first chance to take a break from the direct blast of the wind. I'd come to a corner, the intersection of four fields. There grew a total of four willow trees, all lined up. I unrolled my bundle, carefully pocketing the precious tinder box, and sat on the piece of sheepskin I'd draped over a protruding root. My survival experiment had degenerated into a real situation. It was time to assess.

If I continued on to the woodlot as planned, it had become obvious that I wouldn't get there before dark. Plus, I would surely be dead tired, "toasted on both sides" as outdoorsmen say, and no doubt too close to deadly hypothermia for comfort. If I went back, my situation would be even worse; I'd be facing almost directly into the wind, I didn't know if I could find a way to cross the ditch again without getting wet up to the waist, and even if I could find my way back to the bridge it would be a long walk along the gravel road to the first house in this deep snow. In such a predicament, I would counsel my students to "inventory your possessions and the environmental resources!" So I did. Apart from my short equipment list, I had tons of fluffy snow and four trees, the biggest one of which offered partial shelter from the wind. From those trees, the odd dead sucker shoots offered a bit of firewood, but barely enough to keep a fire going for an hour or two — certainly not overnight. How did I get myself into this mess?

As I shivered while contemplating the three ugly options, my stiffening pants enticed me to decide in favour of starting a fire right then and there. I cleared away the snow and started whittling some fuzz sticks from a dead branch. Soon the flint and steel coaxed them alive into dancing flames. The show was spectacular, the wind tossing those flames into somersaults. My smoky eyes shed forced tears of appreciation. At least my pants were drying!

In what seemed like mere minutes, the wood was gone and my eyes and lungs were sore due to the twirling smoke. But I was dry and I'd had a welcome drink of melted snow. I followed my tracks back to each of the willow trees to see if I could muster up more wood. After some prying, I managed to rip out one of the leg-sized sucker shoots that had previously resisted my efforts. Several other pieces of fuel fell prey to my eagerness.

As darkness fell, a last ditch effort to scrounge up every possible piece of firewood became necessary. I knotted my two thongs together and lengthened the improvised rope with my shoelaces and belt. Armed with this mini-lasso, a couple of low-lying branches were added onto the chaotic collection of firewood.

The pile remained tiny. There was no way it could keep a fire going for more than two hours, I estimated. Soft willow burns like paper, especially with a fierce wind fanning it as in a forge. Then what? As the last bit of daylight disappeared, I pondered why I was there, suffering like I was. Am I nuts? Crazy? I tried to convince myself that my experimentation was worthwhile, with minimal success. Then I took refuge in the fact that I'd always fed upon adversity. Like the old saying goes: "What doesn't kill you makes you stronger!" Though I didn't like the alternative the saying suggested....

With no choice but to spend the night, an idea emerged. I divided the wood into four equal heaps, which I reserved for half-hour fires at ten o'clock, midnight, two o'clock, and four in the morning. That way I figured I would be able to warm up a bit between stints of light calisthenics. I used one of the piles as a seat and two others I leaned on each side of the tree to break the wind, with limited success. The last pile I broke into manageable pieces and set-up tee-pee fashion. I reserved the driest and straightest piece to whittle into shavings.

Reality hit me hard: the drafts made me feel like I was dressed in a sieve. It was physical and mental torment just waiting until ten. And when the long-awaited hour finally arrived, the fire's smoke just switched the type of pain from biting cold to eye and lung irritation. Agony. I couldn't even manage to cook the rice I was looking forward to. The fire was simply too unmanageable and short-lived. Warming my bare hands on the remaining coals, I attempted to use the sticky snow from near the fire to widen my lone tree wind-break. Useless. I finally stomped out what was left of the embers to obtain the relief of a smoke-free environment. Pitch black and bitter cold once more.

It seemed like forever until midnight, and longer still until two o'clock. Excruciating. I couldn't believe I *volunteered* for this. It was absolutely the worst night of my life. If I didn't freeze to death, I swore the smoke would finish me off anyway. It was the wind lesson all over again. Definitely, unquestionably, categorically, undeniably, and without doubt, avoiding wind would forever be my absolute number one priority in any cold weather survival situation!

At four o'clock I entered the smoky torture chamber to get my eyes whipped for the last time. I was burning the wood underneath me, which had permitted lying in fetal position, and I was back to sitting on the crooked root. Nevertheless, the miracle of fire amazed me as always, and my toes welcomed the heat. In spite of an exhausted state, I felt overwhelmed by the grandiose spectacle of nature that surrounded me. Awe inspiring. What a privilege to witness first-hand how minuscule a lone man can be in the face of such power!

As if to acknowledge my revelation, the mean wind softened somewhat. I shivered until morning, but it didn't matter anymore. I knew I would survive. As daylight broke on André the farmer lost in his field, some type of strange calmness overcame me. Perhaps to match the same calmness around me, for the wind had died down. The storm was over.

I reached for my eyeglasses, safely stored in the pot. My vision remained somewhat blurry. I wiped the specs with a corner of my shirt, to no avail. Then I realized I must be suffering from mild corneal abrasion.

Another close call. If more firewood had been available, I would surely have been half blind that morning. And perhaps also severely ill from carbon monoxide poisoning.

After a few minutes of involuntary tears, I rolled up my bundle and tossed it onto my shoulder, happy to leave the horror show behind. In the far distance, I could guess the presence of the woods I was headed toward. Typical of Colorado's instant weather changes, the sunrise pierced the few remaining clouds. It would be a nice day. My spirits lifted with the increasing luminosity, which highlighted the pretty blanket of fresh virgin snow. I savoured the spellbinding white beauty; a gift to compensate the previous night's hardships. Nature's dichotomy never ceases to amaze me.

I wove around the drifts, seeking the shallowest passage to my destination. Two harsh hours later, I finally arrived and immediately sought the comfort of a warm and cheerful fire — in a dense wind-free location. As the wild rice boiled, I sacrificed my leather thongs and started fabricating a pair of quick and dirty snowshoes in order to avoid having to dry my pants and boots each time I drifted away from the campsite.

After ingesting half my daily ration of calories, it was time to gather a substantial pile of firewood. Once that task was out of the way, I found two standing rotten trees of a fair size, which I pushed over and dragged back to the fire spot. Placed side by side a metre away from the heat, they made a reasonable bed for me to crash on. How sweet sleep is when one is so exceedingly exhausted!

My slumber was deep, interrupted only by the sporadic half-awake tug on the long logs to advance their sooty ends back into the fire's heart. A few hours later, I was a new man. Well, comparatively new. Under such circumstances, the lack of comfort has always woken me too early, as if by an unwanted alarm clock.

I headed out for more firewood. When I thought I had enough for the night, I doubled the amount, and then doubled it again. My rule is twenty logs as big as I can carry per night, at least. After I accumulated a two-day supply, I made my way to the river's edge where I wandered, on the lookout for soft and dry bedding material such as goldenrod or fireweed. Suddenly, a floppy object attracted my attention. Wow, it's a lame duck! Supper! I pounced on it with the instinct of a wild cheetah, quickly putting an end to its misery. I've always hated killing animals, but I consider it an act of necessary evil. Nevertheless, I briefly regretted my decision. But my grumbling stomach convinced me to think otherwise. As I examined the bird, I could feel the shotgun ammo that brought it down. Some hunter's bad luck. My good fortune.

I snowshoed downstream along the Platte River to the confluence, the mallard dangling from my belt, and started my way back to camp by going upstream along the Cache la Poudre River with the intention of circumnavigating the woods back to the spot where I had entered it. After a few hundred steps, I was stopped in my tracks by a shrill whistle that came from behind me. I turned around and went back to the point at the confluence to find Bill yelling from across the river.

"Wildman, are you okay? I was worried because of the storm. It was the worst in Colorado history!"

"I'm fine Bill. Thanks for checking up on me!"

"I see you made yourself some snowshoes. Should have known. Where the hell did you get that duck?"

My rule is twenty logs per night.

"Caught it with my bare hands." I didn't tell him it was lame, of course. There goes my ego again, trying to impress.

"Unbelievable! You're da crazy Frenchman!"

After some more small talk, since he announced the weather would be fine all week, we agreed that he would pick me up in five days at the spot where he dropped me off. Then he left as quickly as he came, a flabbergasted spectator in front of the magician who survived the storm of the century without the flinch of an eye. If he only knew of the pain behind the trick. His departure left me proud of my performance, yet so lonely in the knowledge of my deceit. It's a tough price to pay for ownership of that super ego of mine.

As I sat calmly later that evening, toasty warm between the fire and the huge overturned log behind me, delighting in roast duck and wild rice, I thought of my dear old mom back in Scarborough, thousands of kilometres away, who in the face of adversity always maintains that after rain comes sunshine. You're right, Mom!

The South Platte River.

The Farmer

The rest of the week was rather uneventful, other than the loss of a couple of kilos due to short rations. The snow had completely melted away, and clement weather made life relatively easy, permitting me to sleep in the daytime sun without tending the fire. To allay the loneliness and boredom that takes over during such waiting stints, the search for food becomes the centre of attention. On this trip I tried every trick I knew to obtain food, but the luck of the draw was one duck, period. No fish nor game would fall prey to my lines and traps, no significant edible plants could be unearthed. Nature decides. As far as bugs and worms were concerned, my thousand calorie per day diet was sufficient to eradicate all desire to experiment.

The piece of sheepskin I slept on brings back amusing memories. In an anthropology class, I had convinced the professor that it would be useful for me to learn to sew furs and she agreed to my special project of confectioning a sheepskin parka. Tanned sheepskins were too pricey for me, but a local sheep farm could supply raw sheepskins for next to nothing. I purchased twenty of them over the phone, which I decided I would tan myself. Easily said but not done. The wool was so dirty, disgustingly so, and the other side on which meat and fat remained stuck was no better. I spent two full days scraping those skins over an inclined beam made from scrap lumber, using a large axe file onto which I had taped some handles. Then I soaked them in the boarding house bathtub and washed them with laundry soap and borax.

Meanwhile, I was off to the library to find tanning recipes, and found one using sulfuric acid, easily obtained from an old car battery. While tanning the skins, the acid ate right through the bathtub enamel, leaving me less than popular with the roommates. And the stench from boiling scraps of skin — my first attempt at making hide glue — sure didn't help.

The recipe then called for washing the skins in gasoline to prevent bugs from attacking them, then to rub them all with hot cornmeal to remove the gasoline odour. The next step was to hang them on the clothesline and beat them for hours to get rid of the cornmeal. The hardest part was scraping them back and forth over an inverted dulled axe head, steadied in a vice, for the entire time they were drying. A procedure that took over thirty hours straight, and I had no choice but to work at it hard, otherwise the leather would become stiff as a board. I ended up with a dozen useful furs, the rest having hardened following my inability to keep up the pace. That was the first time I regretted my ideas of grandeur, and admitted that maybe I had exaggerated in selecting the number of skins for my first attempt at tanning. After some sleep, I had to rub down the good skins with lanolin to keep them soft and then remove the excess lanolin by absorbing it with piles of sawdust. Another trip to the clothesline for a broom-beating session and I thought I was done. Until I started the combing process, which took another several days.

Hanging the sheepskins to dry.

A month later, after three hundred and fifty hours of sewing each stitch with pliers, I had created an incredible parka of Inuit design, complete with hood, which earned me my course grade. The garment was so thick it was way too warm to wear, no matter the weather. After lugging the parka around for a couple of years, I finally cut it down into a vest, which I wore on winter camping trips for a long time to come. One sleeve was converted into the sleeping pad I used on the adventure I just related.

Nine

Sandals

"I felt sorry because I had no shoes, then I met a man who had no feet."
— Persian proverb

When I first came to Colorado, I couldn't believe the number of flats I was getting as I commuted by bicycle to the university. The culprit was the Southern Sandspur, *Cenchrus echinatus*, whose sharp and tough spines puncture tires like a sewing needle popping a balloon. I was to become intimately acquainted with this special plant.

Here I go again, keyed up by yet another survival experiment. Enough fooling around. Time to go naked! Well, okay, almost naked — I'll wear a bathing suit, just to preserve my dignity, and perhaps also to avoid being shot by spooked locals. (Hey, I'm a stranger to America, after all!) This will be my ultimate test, the one I've been aiming for all these years. Every time I go out to practise, I like to set a challenge for myself that honestly feels like a 50 percent chance of failure. The goal is to propel good old me into survival mode — on edge and apprehensive. In this case, nothing but a bathing suit sounded about right, since this September weekend announced warm days without bugs nor rain, and cold nights just above the freezing point.

I shouldered a small pack with a double meal, hopped on the bike and pedalled off toward the east for a dozen kilometres or so until I reached Colorado County Road 388, which parallels the South Platte River. At some random point I picked an inviting unmarked trail that led to the water's edge. It was a beautiful day and I appreciated my early lunch on the sandy shore, psyching myself up for the task at hand. Enough procrastination, time to strip down. Yes.

As I swam across the river, I left my pack and bike behind both physically and mentally. In my mind I had become a primitive man, naked in the wilderness. To my stubborn brain, this was real: no turning back. I reached the opposite shore easily, but anxiously. Game on!

Stepping out of the water, I was greeted by immediate pain in the form of a sharp prick in the sole of my right foot. I let out a yelp as I instinctively hopped onto the other foot, only to shriek louder still. It too suffered the same shooting pain, and as I danced from foot to foot and from heel to toe, each movement was met with debilitating and instant stings. Airlifted by adrenaline, I managed to bolt over to an overturned log and sit. Then I started picking the dozen or so sandspurs from the soles of my feet, like so many thumbtacks stuck in a corkboard.

A careful scrutiny of the ground around me confirmed my fears — the soil was literally plastered with the spiny burs; no doubt dumped there in the spring by the current at the bend in the river. Buggers! I was stuck on my log. I sat there immobile for a while, in the shade of a bush to avoid sunburn, wondering what I was going to do. I tried to sweep the ground with a stick. Not effective, I'm afraid. Now what?

The log I was on was an entire tree that extended to a considerable length. I walked along it back and forth, looking for options. At one end grew a willow shrub, and I started peeling its shoots with my nails to obtain long strips of bark. I fashioned them into a twine using the inverted double-twist technique.

After a while I had several long strands, which gave me an idea. I broke eight or nine foot-long pieces of green twigs of pencil diameter and lashed them side-by-side to construct a miniature raft that would act as a shoe sole. To that I attached a willow bark braid in a manner I learned in Mexico, where artisans produce car tire flip-flops. After repeating the

Sandals

The inverted double twist technique for making cordage from natural fibers.

process, I was the proud owner of a pair of primitive sandals. I treaded carefully at first, but with use gained faith in my improvised footwear.

By then the afternoon had largely been consumed, and I still needed a fire for the night. It wouldn't be an easy task without tools, my record thus far being just over two-and-a-half hours. I found a few rocks that I broke to obtain sharp edges and started looking for the parts needed for a bow-drill fire: straight spindle, split board, handle, and bow. In that type of prairie forest where cedar or basswood are absent, willow wood is the most appropriate for the task. Without metal tools, it wasn't easy to sharpen the spindle to precise points, something I had to accomplish by grinding the ends on a solid rock. The board was easier to obtain, I simply broke a branch by wedging it between two adjacent trees. For the handle, another piece of broken willow would have to do; there was no harder wood to be found. Without grease in the handle, the friction was doubled, the only partial solution being to narrow the top of the spindle to pencil thickness. More rubbing! At least everything was bone dry.

I finally wrapped the spindle around the twisted willow-bark rope of the bow and started the seesaw motion. A curl of smoke rose from the board. Once enough wood had been scratched by the drill to define the hole, I stopped and ripped a notch right through to its centre by sawing with a rock flake. I resumed the familiar back and forth bow motion, letting the notch fill with dark brown powder. Then I accelerated and pushed harder on the handle to raise the heat in the wood dust to combustion temperature. But the notch was cut too wide and the spindle fell into it. My record wouldn't be broken this time! I started over, but just before success, the narrowed section at the top of the drill wore down, creating friction against the handle. With this extra resistance, I didn't have the power to spin the drill fast enough to generate an ember. My shaky arms once again ground down the spindle's diameter by rubbing on a rock.

I resumed spinning, and smoke poured out. But the drill had become too short and I couldn't hold it steady enough. Failure again. I had to build a new drill from scratch. On the next try, the weak willow rope broke for the fifth time. Darn! I thought I needed a larger diameter spindle to ease the wear and tear on it. Frustrated, I cheated and pulled out the string holding up my bathing suit and tied it to the bow. There was nobody there to comment on my plumber's crack anyway!

Finally, a shout of glee welcomed the glowing ember. With the utmost care, I transferred it to a nest of dry grass and blew it into precious flame. Now I could survive the night, but all the delays pressed me for time. The evening shadows were lengthening by the minute as I scrambled to gather firewood. Soon the cold set in, and I had to warm myself by the fire in between each run for fuel. Predictable blisters formed where the sandals were rubbing my feet raw, but I couldn't stop yet. Dry cattail stalks and leaves were the only material I saw soft enough to lie on, and I would need armfuls and armfuls to make my bed. I gathered a very huge pile until it was too dark to see, and also pulled up several of the stringy roots for supper. At last, I could collapse on the pile of stalks and remove those dreadful sandals.

I was dying of thirst and a terrible headache indicated the first stage of dehydration. The South Platte passes through town and is obviously too polluted to drink without boiling. In preparation of my need for a pot, I had placed a huge dry log in the fire to cut it into a manageable

Sandals

metre-long length. Onto this log I transferred a bunch of coals with a bark shovel and fanned them with the same instrument. In time, the coals burned their way down into the log a little. With a piece of rock I scraped the burned part off and transferred new coals to the depression thus formed. By repeating this process over and over again until past midnight, I obtained a very crude wooden bowl. My respect for pirogue-making Indians had climbed a notch.

By moonlight I carried the log to the river and filled the depression with almost a litre of water. With two forked sticks used as tongs, I transferred a fist-sized red-hot rock into the water and watched it sizzle. With a second rock, the water steamed and was purified, albeit a dirty ash black. But the camel needed a drink badly and appreciated the soot tea. He requested a second serving, then a third.

Even with a bright fire roasting my front, my back remained as cold as an ice cube and I had to continuously rotate. I was starting to feel like a barbecued chicken. Impossible to sleep. I decided to make myself a blanket with some of the cattail stalks and leaves. I'd done this before, by tying bundles of material together, knotting them with braided cattail

A cattail blanket I made on another occasion.

leaves. From two o'clock until morning I finally managed to get a bit of shut-eye, in half-hour stretches at a time. In between, while re-arranging the logs in the fire, I sang: "My dad is rough, my mom is tough, and me I'm rough and tough."

As can be imagined, I didn't sleep in very late the next dawn. I think I swam across the river before the sun was up and dove into my dry clothing. To celebrate the end of such an outing, there's nothing quite like a couple of peanut butter and jelly sandwiches with a liter of juice. Here's a toast to you, Mother Nature! Thanks for letting me in.

After riding home and sleeping in a warm bed for twenty-four hours straight, I was none the worse for wear. On the contrary — I felt alive.

Ten

Licking My Chops

> "So long as you have food in your mouth, you have solved all questions for the time being."
> — Franz Kafka

Let me introduce my younger brother Michel: even more of a nut case than I am, believe it or not. You bet Michel a dollar he can't do it and he'll guzzle down a whole jar of pickle juice. Another dollar and you can get him to run barefoot to the other shore of an icy lake and back. At two hundred and fifty pounds, he's somewhat chubby, yet strong as an ox. And perhaps more importantly, tough as nails. We tease him: "No brain, no pain!" and he laughs. Nice guy.

Michel is one of seven recruits I somehow convinced to enlist for this weird trip I conjured up — rule-bound winter camping for the nine days of March break, which I would spend in Ontario as a welcome escape from graduate school. My youngest brother Réal was also on the trip, a reasonable young man who didn't know what he was getting into, as well as my cousin Jean-Pierre and four of my former outing club members. Our challenge seemed simple enough. Since we carried our warm sleeping bags we would only have to build a group shelter. Oh, and the sole constraint imposed the choice of a single basic food item for each person, no trading or sharing allowed. I figured we'd be so bored by the repetitive grub that we would be inspired to scrounge for wild food.

After much consideration, I filled my own pack with a ten-kilogram bag of rice. Plenty. I figured the millions of Asians who survive daily on rice can't all be wrong; plus, I imagined rice cakes, puffed rice, and rice cereal. Réal went with flour; as a baker he was thinking of tasty flatbread. Jean-Pierre loves macaroni noodles so his choice was easy. I wondered what Michel and the others had selected.

The restaurant lunch stop was the highlight of the long boring drive to the edge of the woods. As I pulled the bags out of the station wagons, I was astounded by the weight of Michel's pack. He sure wouldn't run out of food! We donned our snowshoes and followed Jean-Pierre's lead into the forest, single file, *à la queue leu leu*.

An hour and a half later we decided to establish our camp near a random wilderness lake. We spent the rest of the day constructing a mammoth evergreen lean-to in which all eight of us could slouch side-by-side, feet kissing the toasty fire. It was pitch black before we all got around to cooking up our individual fare: two rice, three flour, one macaroni, one egg noodle. Michel, woozily stretched out inside his sleeping bag, looked and acted like Garfield the lazy cat.

"Aren't you eating, Michel?"

"Nah. Couldn't be bothered!"

The rest of us gulped down our boring grub and hit the sack too. Rice without salt sure is bland! In spite of the minus twenty degrees Celsius, all of us woke to a sunny morning completely refreshed. Except Garfield Michel, still snoring away. Soon rice, pasta, and flatbread were cooked up again, but none of us were enthused by the contents of our respective mess tins. It hit me how our happiness revolves around food. On regular trips it's the centre of attention, that's for sure.

I headed out with a couple of others to replenish our dwindling firewood supply. The crisp air filled my lungs and spirits with that wonderful sensation we went there to seek. I soaked up the sun, enjoying its pure energy. Wood was plentiful. I lined up several long pieces into two bundles on each side of me and tied their butts with the opposite ends of a thick rope, leaving a metre-long length between. With this improvised harness draped behind my neck and around the front of my shoulders, I dragged the load back to camp.

As the shelter came into view, my nostrils detected an ecstatic whiff. It smelled like charcoal steak! Then I saw it. Michel was sitting by the fire, roasting hunks of meat he had skewered onto a sharpened stick, shish-ka-bob style. On a log beside him sat a monstrous entire top round of beef that surely weighed twenty kilos. Man did that seem tasty! I was licking my chops just looking at it. I stared for a while at those mouth-watering morsels grilling on the coals. Then I noticed my other deprived companions, all of them salivating like a bunch of hungry dogs at Pavlov's beef. While no-brain-no-pain Michel growled out slurping noises, grinning from ear to ear.

Of all my lifetime survival trips, I swear the next few days were the most painful of them all. As the boredom of rice increased, the sight and smell of Michel's fat meaty marshmallows became absolutely intolerable. All of us had to head out for a long walk three times a day to preserve our sanity. To waste time, Jean-Pierre carved decorations onto his new walking stick, some played checkers with a made-on-the-spot set, others created artwork by scratching the surface of *Fomes* mushrooms, and many an elderberry whistle was blown. We tried munching on basswood buds, spruce gum, and rock tripe, but this fare paled in comparison to the food in our possession, even if uninspiring. We also profited from the rare occasion of skating-rink-like ice on the lake to engage in improvised hockey games using homemade sticks and a wooden puck made by sawing a slice from a maple branch. The exercise compounded our desire for caloric-rich food and we envied Michel more and more.

We were saved from the look-at-the-steak torture at the end of day four. As I stood on the lake in an unsuccessful attempt at fishing, I observed some circling ravens far away in the distance. I hiked to the end of the lake, over a small mountain to another lake, then over a hill to a pond above which the ravens were cawing. After some patient waiting and searching, I came upon three skinless beaver carcasses that had obviously been left there by a trapper. They were frozen solid, and one of them had not been pecked at too severely. I chopped off the damaged meat with my hatchet and shouldered the rest of the animal to bring it back to camp. After the beaver thawed by the fire we all accompanied Michel in a shish-kabob feast that evening. I couldn't help but wonder if my buddies would have followed me on this adventure had they known they were to partake in crow leftovers!

The next day, wanting a bit of brotherly revenge, I challenged Michel to an overnighter without gear other than an axe. He accepted with unsuspected enthusiasm. The night hovered at below twenty degrees Celsius; the challenge was tough. I came back into camp the next morning severely beaten up. Surviving the night wasn't particularly problematic, but it sure was a lot of work. The first task was to use a snowshoe to dig — all the way down to the ground — a metre-deep trench parallel to the wind, more than ten metres long by one metre wide, in the centre of which I started the fire. The trench served to oxygenate the fire and I had previously learned the hard way that without it life is nothing but smoky misery. Next, I placed the ends of a dozen or more full-length logs onto the emerging flames to produce a bonfire. As I went to gather tons more firewood, the heat from the bonfire melted the snow in front of it, and I was then able to enlarge the living space thus formed, building up walls with the softened sticky snow. The traditional metre-thick bed of long evergreen branches, created by poking their butts into the snow in forty-five degree angle rows, completed the setup. In spite of the relatively calm night and abundant firewood, the outcome was nevertheless twelve chilly hours of tossing-and-turning intermixed with laborious fire maintenance. Come early morning, I was sure glad to crawl into my sleeping bag to catch up on sleep.

I awoke at lunch. To my surprise, Michel seemed a little ruffled, but none the worse for wear. I decided to follow his tracks to check out his previous night's campsite. The tracks led to another lake, much further away in the direction opposite to where I had been exploring. At the end of that lake, in a hidden bay, his tracks ended at an unlocked cabin. The stove was still warm.

Michel was laughing so hard when I returned that he felt obliged to accept the challenge for real that night. As he stumbled into camp early the next morning he wasn't amused anymore. He was dirty with soot from head to toe and looked like he had been run over by a cement truck. Poor Michel!

Eleven

Batman

"Not until we are lost do we begin to understand ourselves."
— Henry David Thoreau

"André, you like challenge, why don't you come with us on a spelunking adventure?"

"What's spelunking?"

"It's the term we use when we explore caves. It's a synonym for caving."

"Sounds like fun to me!"

"Loads of fun. Especially since we've discovered a new cave which hasn't been fully explored yet!"

"Wow! Where is it?"

"In Missouri, not too far from my parents' house. I know a couple of great guys in our spelunking club there. They're experts. Plus they've got all the equipment we're going to need."

Missouri is quite a ways from Colorado, but my buddy and fellow outdoor student Chuck knew I was heading back to Canada at the end of the spring term and hoped I could make it a stopover visit. His desire to hitch a ride with me must at least have partially motivated his offer. That was okay with me, since I was totally thrilled and energized by the prospect of trying a different outdoor activity, of discovering another type of wilderness.

Three weeks, a bumpy Jeep ride, a wilderness campout, and a brisk early-morning hike later, we were peering over the edge of the sinkhole where we would descend. But first we distributed the group gear, which consisted of a variety of ropes with slings, carabiners, and other climbing paraphernalia. Oh yes, and a pot of grease, I didn't know why. Due to weight considerations our personal gear was limited to a lunch, some extra clothing, and, most importantly, our hardhat helmets mounted with carbide lamps and extra calcium carbide fuel. We were wearing raingear as an outer layer, over wool clothing of course.

Climbing into the abyss, or what seemed like an abyss to my uninitiated mind, we ended up on a ledge two stories below. This happened to be my very first contact with a sizeable cave and impressed is a small word to describe my feeling. I blindly followed the leaders obliquely down and then through a maze of passageways, too overwhelmed by the incredible stalagmite-stalactite oozy environment to think. They had been there more than twenty times and had memorized the layout, so I trusted them. After half an hour or so we entered a lone passageway and hiked along it for the longest time, unconcerned with direction finding. I was amazed. The variety of shapes and forms as the passageway swelled and narrowed spellbound me. This must be a dried up underground river. My experienced buddies confirmed.

Eventually we came to a cavern the size of a living room. A few bats hovered, emitting their weird communication shrieks. For an instant, I was Batman in his bat cave. But the movies can't show the stench: the air felt stale, stuffy, musty, and rotten. The grotto seemed without issue, and I wondered why the biggest fellow amongst us was undressing. My query was quickly answered by a finger pointing to the tiny but man-sized hole I hadn't noticed. Apparently the passageway continued beyond this obstacle — that was the discovery they had been referring to. I watched in fascination as the bulky guy smeared his body with grease from the pot, and painfully wiggled to squeeze through the narrow opening. Small me crawled through effortlessly.

Beyond, the channel continued for another kilometer or so, then divided into two branches. We veered to the left and travelled downhill. Soon we were wading knee-deep in freezing water but continued the exploration. In one spot a narrow deep-water pool forced us to cross above it while crabbing sideways in push-up position, hands and feet on opposite walls. We crossed several musty-aired chambers, some of which were fair-sized, weaving around stalactite columns. Sometimes we crouched low, but thankfully we could stand most of the time. We came to a vertical fork. I was told the downward leg came to a dead end, so we climbed up with ropes to another level where something special apparently awaited us.

A few hundred metres forward along the new corridor and we'd arrived. I was awestruck by the incredibly colossal cave we'd just penetrated. It was the size of a gymnasium! Our voices echoed repeatedly as we expressed our wonder. Wow! We refilled and adjusted our carbide lamps. The spelunking club had been this far before, but the nooks and crannies were still virgin territory and we were free to explore.

Hunger interrupted our investigation. It was way past lunchtime and my hefty roast beef and cheese sandwich received an envious look from my granola-bar-addict friends. My cold feet appreciated the plastic-bag-covered dry socks I changed into. We pursued our inspection of the cave, scrutinizing each recess in the hopes of finding further issues, but in the end we had to admit defeat. In any case, after seven hours underground, everyone was looking forward to fresh air.

We made our way back to the squeeze hole then started the hike back to the entrance. I felt great, privileged to have witnessed such a grandiose act of nature. Like a horse scurrying back to the stable, I pushed ahead, comforted by the chatter of my buddies behind me and the occasional presence of our former tracks in the moist soil. It was quite a long walk. I fell into a daze, as one does when long-distance canoeing or hiking, dreaming of adventures to come. The repetitious rocky scenery melted into a blur as I whistled along, losing track of time. Superb.

Then I abruptly awoke from my reverie to eerie silence. No more voices behind me. I yelled. No answer. I slowed down to let my buddies catch up, as the passageway climbed and narrowed. Then I hit a dead end; I must have somehow missed a turnoff. I went back the way I came, suddenly aware that the hard ground there left no footprints.

After a few minutes of backtracking, a familiar chamber appeared. Off to one side I found the channel I missed, up higher. Wait, there was also another tiny passage on the other side too! How could I have missed these? I climbed up the most promising of the two exits with the hope of remembering my way. But like the true beginner I was, every stalactite, stalagmite, and rock face looked identical, as must seem every tree in the forest from the perspective of city folk. To make matters worse, the passage I was in split into two forks. I followed the upward one, which diverged again. I ended up in another chamber, with the only way out being to climb up or down. Instead, I went back, but encountered a reversed fork and in a moment of panic couldn't remember which one to take. I guessed the left.

Wrong choice. I ended up back where I started, but up one level. There seemed to be passageways in every direction including up and down — a three dimensional maze. My survival instinct said start a fire. Stupid thought, no wood there. Then I pondered waiting it out until morning light, and panic hit hard when I realized that there would be no sunrise in those depths. I panicked even more when I estimated that my carbide lamp contained fuel for no more than two hours of brightness. Then I would be left with the candle stub I shoved into my pocket prior to departure, and a container of waterproof matches. Fright intensified and I ran around some more. I was hopelessly lost. Cinematic visions of search crews finding my whitewashed and faded skeleton appeared. Gulp!

A flashback to my own lecture of what to do when lost saved me from further panic. *STOP! Mark the spot where you are right now. Don't lose this spot!* The reasoning is that from that precious point you are close to where you were when you weren't lost and you won't worsen your situation. In the woods I suggest marking the spot by building a fire or a huge tripod, or by fixing a log horizontally between tree forks. Then I tell my students to leave their useless-for-survival money or credit card there, as an incentive to not lose the spot. In the cave, the best I could do was to pile rocks into a mound.

Satisfied with my cairn, I picked an alley at random and checked it out, scratching arrows into the ground to indicate the direction back to my mound. The path lead to a dead end. I returned to my mound and used more rocks to form an X to prevent me from returning along that particular cul-de-sac. The next direction I took led to a chamber in which the stalactite-stalagmite pillars were so tight I could hardly squeeze through. Since I'd never crossed such a chamber, I put an X at that path too. Two more conduits faded out into dead ends. Then I climbed and took the next available corridor, abandoning it when I had to crawl. My mound of rocks felt like home base, and I sure was glad to touch it again. A few X's later, I was relieved to find an arrow scratched onto the wall of a chamber, pointing upwards. The face was too steep to climb, and since we never rappelled down such a face, I figured there must be another way up. I scrutinize a few other passages, never losing my rock mound, but was getting desperate. I returned to the arrow, just glad to see a human sign.

As I examined the rock face from a climber's perspective, I saw a glimmer of hope in an upward-running crack. The ten-metre high climb seemed treacherous, but I decided to attempt it. Some cold sweat and I was at the top. But there was no way to climb back down safely. Worried sick, I took the widest passage up an incline. A brief minute later, I was admiring the sunset with the boys.

When I exited that cave after having spent twelve hours underground, my friends were nonchalantly hanging their wet clothes up to dry on nearby bushes, beers in hand. Were they worried? Not one bit. After all, they had arrived but half an hour ago, and they assumed I merely wandered off into the forest to answer nature's call. When I narrated my misadventure and climb out, they just shrugged and told me I should have gone around the other way. Apparently I had missed the regular exit. It turned out I was in no real danger, since this end of the underground complex had no other issues. I figure I had outpaced my friends by half an hour, which means I was lost for no more than an hour. The longest hour of my life.

This event taught me a great deal about survival psychology. I had evolved for so many years in the forest that it became familiar and I no longer faced the unknown. And it is the unknown we fear, which is often nothing but a perceived ghost overshadowing the simple reality of the situation. The dark scary face of this strange wilderness had thrown me for a loop.

Twelve

Beans

"*The most difficult thing is the decision to act, the rest is merely tenacity. The fears are paper tigers. You can do anything you decide to do. You can act to change and control your life; and the procedure, the process is its own reward.*"
— Amelia Earhart

"Michel, stop feeding what's-her-name that cookie! She'll crap in the van again!"

"Her name is Paulette and she's hungry, poor thing."

Paulette was one of three old hens we'd purchased from a farmer, half of our food for the four-day survival experiment we were hurling ourselves toward. The other half consisted of a bag of navy beans. The challenge this time? No container to cook in. And winter camping with only two summer sleeping bags and a pup tent between the three of us, the third Stooge being our friend and former outing club student, Bert. This guy was sharp, the brightest student I had encountered as a teacher. One day for example, I had showed him how to boil an egg in a paper bag and the next he had come back having duplicated the feat in a burdock leaf. Another time I showed him to make toy animals with magician balloons, only to be subsequently delighted by his perfect Adam and Eve interpretations. He was the perfect candidate with which to share our folly.

We cut our anxiety during the drive toward this fifty-fifty confrontation with nature by joking: first about how Michel wouldn't be able to wring his chicken's neck after bonding with her, then by recalling Michel's gag "zonk" gifts he gave everyone at Christmas Eve two days ago — penny-and-paper-clip earrings for our sister for instance — and finally by reminiscing teasingly about Michel's tire-screeching adventures in his junkyard-parts souped-up 440-magnum Franken-car. Oh yes, we also chuckled when we harked back to the time he carved the spit roast with his chainsaw.

I ended the dull part of the journey by parking the van on a straight stretch of the now familiar Bear Lake Road. Bert and I grabbed our packs and our feathered damsels, and snowshoed off into the woods, breaking trail. Michel immediately followed, with Paulette on a leash. Bert and I were cramped with laughter as he tried to tenderly coax her along.

It was bitingly cold when we reached a promising camping site at the base of a cliff. As planned, Bert and I endeavoured to use deadfall traps baited with the cookie to slay our animals, but the hens and Michel refused to cooperate. We therefore switched our attention to digging the usual fire trench, setting the tent on a thick bed of boughs with the front door opened to let in the blaze's heat. The el-cheapo sleeping bags sure looked skinny once laid out on the floor of the petite shelter.

Near the end of the day, to Michel's mostly faked chagrin, the cold sent Paulette to the great chicken-coop beyond. This, plus hunger, incited Bert and me to become poultry executioners. By some strange twist of fate, the two headless birds fluttered directly toward Michel. The big tough guy became the proverbial elephant scared by a mouse and added humour by acting the part. Tears of hilarity thawed our cheeks and relieved somewhat the slaughterhouse atmosphere. As we plucked and cleaned the birds with frozen fingers, we couldn't help but reflect on how far removed from reality chicken-nugget-eating city-folk have become. And how the fruit of our labour can be so easily purchased, leaving the illusion that we are not responsible for the animal's demise.

All of that was forgotten once the fake partridge, roasting skewered over the glowing embers, glazed to a golden brown. We found the meal stringy yet delicious, but one lone fowl does not satisfy the appetite of three hungry bushmen. I was the only one who appreciated the desert of giblet shish-kabob. While I was busy licking my fingers and Bert busier still fetching more wood, Michel erected a cross over the buried mound of leftover bones, complete with a charcoaled-inscribed R.I.P. plank. We chuckled, starting to wonder if he was serious.

The bright full moon rose to fade the crisp clear stars, obviously announcing a cold spell that would plummet the temperatures to at least minus thirty degrees. Our faces stung as soon as we distanced ourselves from our blaze. I took the first watch as fire picket, sending my two partners to attempt the clown act of squeezing into the tiny tent. After a couple of hours, Bert exited to replace me, shivering and complaining that the heat did not penetrate sufficiently into the shelter. I replaced him in the sleeping bag, but he was right, it was too cold to sleep for mere mortals. Only Michel snoozed.

But by midnight, even the tough guy joined us and we all squatted intimately near the fire in a semi-futile attempt to warm ourselves. We just existed in a comfortless slouch for a couple of hours, too tired to think straight. Then I reminded myself of my own survival motto: *don't tolerate, activate!* I had to activate my brain and body to at least attempt to find a solution. How about moving the fire away from the cliff wall so we could sit between? A lot of work. Maybe we could build a wall behind us to cut the wind? Hard work too. It wouldn't be easy to fetch building materials in the frigid moonlit darkness. Could we use the tent to erect a vertical wall behind us? I hopped into the tent with the sleeping bag I had draped around my shoulders to examine it. At least in there I could lie down on a soft floor. If only it weren't so glacial! This gave me an idea.

I coerced Bert to hand over the other sleeping bag, and he reluctantly agreed. Luckily they zipped together. Then I suggested that all three of us take off our parkas and try to squeeze together into the double sleeping bag. It was no sooner said than done, and we managed to toss our coats on top of our matrimonial-style bag. With no one left to tend the fire, the tent door was zipped shut, leaving just a small opening for ventilation. The strategy worked; we generated enough body heat to sleep.

It was past two o'clock the next day when our next short order of roast chicken arrived, without French fries. Nor beans. The survival book I recently read made it sound easy to boil water in a birchbark vessel; apparently the water keeps the bark from burning. I was motivated by hunger to work up a side dish of beans.

I'd made lots of birchbark containers before, so I thought it would be a piece of cake. I headed for a suitable *Betula papyrifera*, only to find the bark frozen solid and impossible to remove, except for the wispy thin and fragile dangling layers. So I kept searching until I found a leaning birch tree, under which Bert helped me start a fire. After a while the bark thawed enough so I could peel a suitable piece. To fold the corners of the rectangular container I wanted to create, origami style, I needed to soak the bark in boiling water — duh, that's the objective — so Bert suggested I alternate dipping it in snow and then holding it by the fire. I pinched the creased corners with split maple twigs, making sure to select segments just below a knot to prevent ending up with two pieces. A pot was born, all white inside.

Bert had been making water by melting snow in a t-shirt bag suspended near the fire. The water dripped into the plastic bag in which the navy beans were sold; we had dumped them into a rucksack pocket as part of the let's-maximize-our-resources camping shuffle. I poured the water into my birchbark vessel, satisfied it wouldn't leak. With gloved hands, I deposited it gently but quickly onto the bed of glowing coals, hopeful. Indeed, it was amazing that the bark stayed intact. Below water level, that was. I watched helplessly as the bark above the waterline flamed up, leaving the homemade clothespins hanging until the bark collapsed, with the fire hissing at me as if to shame my ineptitude.

I returned to the leaning tree, where I repeated the process and managed a second sheet of bark. Then I used four flat rock flakes — which had caked off the boulder the fire was leaning against — to install an elevated platform with a square central hole so that the flame would only caress the pot's bottom. I transferred embers under the platform and added kindling wood. For a few minutes my strategy seemed to work, but soon the flames licking the pot's underside poked through the tiny cracks between it and the rocks of the platform, and the second container suffered the same fate as the first.

I searched further in the forest until I found another leaning birch tree underneath which to build yet another fire on a raft of rotten logs placed on the snow. By the end of day, I had obtained three decent sheets of bark, although not as thick as I would like. Failing daylight prevented me from completing my project. We hit our crowded bed hungry.

After a lousy night's sleep, I was up bright and early to face the excruciating cold. I barely had time to strike the match before becoming the victim of frostbite. How glad I was to have prepared tons of birchbark shreds and kindling the night before! Soon a two-metre-diameter warm haven let us melt our frosty whiskers. As my partners attempted to thaw the remaining hen enough to skewer pieces onto a stick, I was back on the cook-the-beans project. I constructed another pot and set it on the platform. With thawed mud scooped out from near the fire pit, I carefully plastered all cracks around the container. Finally, the water started steaming. But the heat softened the too-thin bark, and the container wilted, spilling half the liquid. By then I was more than a bit exasperated by popular survival literature, wondering if the authors had ever abandoned the comfort of their plush La-Z-Boys or Barcaloungers to try the skills they professed.

The partly extinguished fire permitted me to add a couple of fist-sized rocks on each side of the pot and solidify the set-up with additional mud. At last I gladly basked in Michel's and Bert's congratulations at the sight of the feeble litre of boiling water. I tossed in a couple of handfuls of beans. As a chef, I knew that these take a great many hours to cook in liquid thickened with other ingredients, but will cook in as little as a couple of hours if dropped into boiling water only. In this case the process was tedious, as I repeatedly had to add finger-sized firewood under the platform, and occasionally replenish the evaporating water from another container Bert had fashioned. Four hours later, we enjoyed a tiny cup each of flavourless effort. At least it was food, and the setup remained functional for another batch, and another after that. But after having chased firewood all day, we were still starving as we hit the sack, entertained that night by a strange gastrointestinal concert.

On the fourth morning we couldn't stand the masochism anymore, and I was certainly not up to the task of repeating the complex bean routine. My energetic "Let's get out of here!" met with considerable enthusiasm. In no time we had cleaned up, leaving behind no trace but

Boiling water in a birchbark container. Much easier in summer!

a few scattered blackened rocks that would be covered by forest litter within a year or two.

Panting from the brisk walk in a cloud of steamy breath, I've rarely been happier to pat my old van. My partners quickly tossed in their packs on top of mine as I turned the key in the ignition. *Tic.* Again. *Tic. Tic.* Dead battery. *Tic.* Very dead. Crap!

That was the day I learned to leave survival food and gear in my vehicle at all times. Road #12 was at least a dozen kilometre walk, and once there we didn't know how lucky we would be at hitchhiking even if we did encounter traffic. The alternative of staying put was no better: we would no doubt freeze before rescue. I decided to apply the STOP principle and start a fire on the road in front of the truck to warm us up while we would think this one through. After a moment of mindless staring into the roaring flames, it occurred to me that if we waited until the fire fell to a bed of coals we could then push the van over the embers to warm the oil pan and battery; or even scrap our aluminum toboggan or a hub cap for the purpose of sliding embers underneath the motor. This of course teetered on dangerous to me, but were the only options. An hour and a half later, just as we were about to realize our plan, it occurred to me to first try the key, since the bumper was warm. The motor roared.

As we stopped for grub at Dorset village's only and lonely gas station open for business on this holiday evening, we benefited from another valuable lesson: cola and chips don't qualify as real food — we would have preferred more chewy chicken and bland beans!

Thirteen

Banquet Time

"Who wants a world in which the guarantee that we shall not die of starvation entails the risk of dying of boredom?"
— Raoul Vaneigem

The late seventies and early eighties saw me shuttling back and forth from Colorado to Ontario, as I attempted to reconcile graduate work with family obligations related to GB Catering Service. To the despair of my graduate advisor and also my Dad to a lesser extent, I perceived both of these preoccupations as unwanted distractions from my real longing to be exploring the natural world first hand. As soon as either turned his back, I would disappear into the outdoors. In Colorado it was to the mountains I headed, not so much for survival as to hike and climb. Nevertheless, I vividly remember close encounters while stalking elk for fun, or smoking out ground squirrels just to see if it could be done. I also recall picking ticks off my body after camping beneath the stars. I learned a great deal from my experienced friends as I accompanied them on mountaineering hikes, such as our winter ascent of the 4,346-metre Longs Peak. But my climbing career ended abruptly one day as I was ascending the classic Bastille Crack and witnessed another lead climber overhead take a fifteen metre freefall, saved in extremis by the third hex-nut protection after the first two had torn away.

While in the United States, I also appreciated my humbling initiation to whitewater kayaking on the class IV rapids of the mighty Colorado River — which I mostly rode upside down in panic mode. Botany field trips into the Sonoran Desert of Arizona impressed me too, but not nearly as much as the one to Mexico's Chihuahan Desert, my first contact ever with Third World poverty. What an eye opener, both on the human level and because of the new opportunity for experimentation it provided. I found out for myself how unrealistic survival skills such as obtaining water from solar stills or by mashing the inside of a cactus were. In the first case, after sweating profusely while digging a solar still and then waiting all day, it generated less than a quarter cup of water, and in the second, once I had struggled to slice open a cactus and squish its innards, it provided an unpalatable acidic paste with which I could hardly even wet my lips.

Back in Canada, canoe tripping in Algonquin Park and elsewhere quickly became my leisure of choice. I loved bushwhacking my canoe into the most remote lakes, where no one else dared go, where the big bass lurked. What great fun it is at night to catch crayfish by pinning them to the lake bottom with a Y shaped stick — the two tips cut to one centimetre in length — to use as bait for the next day's fishing. And more exciting still to catch water snakes, bullfrogs, or snapping turtles for food, which was legal back then.

Sometimes I would go out for a day paddle and portage over to some unnamed lake to explore. I couldn't help it — as the prospect of discovery pushed me forward I often went too far to return in time before dark. In a way I must have enjoyed the challenge of spending the night by a fire without gear, especially when some unsuspecting friend was accompanying me. Sorry for the cold rough nights chasing firewood I imposed on you guys.

As I was paddling along briskly one late November evening while returning from a multi-day trip in the middle of nowhere, the water gradually thickened with slush, so much so that I was literally stuck in the syrupy soup and couldn't get to shore. Faced with no choice, I pulled out the sleeping bag and slept right there in the canoe. What a memorable night in my improvised million-star hotel. Come morning, I stepped out

Banquet Time

of the canoe onto solid ice, chopped it out with my axe, and finished the trip by dragging the watercraft over the ice to the portage. I was in heaven.

Less sublime were the survival experiments in biting insect season. On one blackfly-infested spring day, having no insect repellent or nets, my partners and I almost gave up within five minutes of departure. "Let's just put up a quick tipi shelter with our ponchos," I suggested while speaking through my handkerchief, unsure of why I wasn't already heading back to the van. Lo and behold, we noticed that in the darkened space the flies would all congregate against the walls and we did manage to spend three days in a smoky environment.

On another occasion, I was bold and stupid enough to face ferocious forest mosquitoes while dressed in nothing but a swimsuit. I stopped running once I found a mud hole and plastered myself with the oozy stuff. To no avail. I was a sitting duck, a perfect Gulliver target for those thousands of Lilliputian darts. I surely broke a record while running back to the van and learned that when Mom Nature says "No," she means "No." That was the only time of my life I had not overcome a challenge I had set for myself. As I sat there in the van swatting at the remaining attackers who had followed me into my fort, with the dilemma of whether or not to get dressed over the caked mud, it occurred to me that I had always won simply because I had set the rules in my favour. When naked, no one can survive minus forty degrees for long.

After this incident, I definitely appreciated winter camping even more. For the rest of my life, every time I encountered cold, I would encourage myself with a loud and enthusiastic "No bugs!" But facing frozen wastes without proper gear remains the ultimate challenge, because it is deadly. Through winter camping, one learns to respect cold. Luckily I have never suffered frostbite, even the time I went with running shoes instead of boots. Or the time I pretended I had burned my boots by the fire and snowshoed back to the truck wearing mitts on my feet.

At this point in my life, I had always conducted experiments that were either short — two or three days — or, if longer, included at least minimal food. I now needed to determine how I would fare without this most fundamental element, and started going on trips without sustenance at all. I quickly discovered that the pleasure of a trip is directly proportional to food intake. And in a survival situation, especially without hunting or trapping gear, calorie count is often determined by pure luck of the draw. On some trips during clement weather, when the blueberries literally coloured my pant legs blue, or when I happened upon a porcupine, or when the smelts were running thick in the creeks, surviving was a piece of cake. But when nature closed shop, it was another story altogether.

I remember one time being so desperate for food I was licking the leftover salty crumbs of pretzels from the last outing mixed with sand on the tent floor. My brain lacking sugar and not thinking straight, I ate some Boletus mushrooms, with but a vague recollection that non-red pored species are all edible except maybe a few that turn blue upon crushing. Which is exactly what happened a minute after I had eaten them. I repeatedly tried to make myself vomit without success, a most unpleasant experience. The fungus did not make me ill, but I sure was sick from worrying.

On most survival trips, my staple food was usually the roots of *Typha latifolia*, the common cattail, which Euell Gibbons aptly named

Gathering cattail roots for food.

the supermarket of the swamps. It's difficult to satisfy one's hunger with this bland and stringy starch, but at least some calories are guaranteed.

But what happens when no cattails dot the scenery? I was about to find out first hand on another survival experiment with my pal Bert.

The super highway between Chicago and Detroit lives up to its name — super boring, that is. I'd been riding my Yamaha 750 motorbike from Colorado non-stop for over eighteen hours. I was tortured by noise. At least I was comfortable. The passenger seat was occupied by my filled-to-the-brim hundred-litre packsack, tightly cinched to the sissy bar with trucker hitches. It must have been illegal, but I installed a homemade cruise-control device, which was nothing more than a flat bar of steel wrapped and squeezed onto the handle just tight enough so it held the throttle when propped against the brake lever. Which meant I could ride no hands, with my arms crossed, leaning back onto my pack with feet up on the crash-bar foot pegs.

My only diversion had been my Teddy, dangling from the ape bars. He moved around a lot, but wasn't too talkative, stuffed as he was. I couldn't wait to see Mom's face, because she wasn't expecting me back so soon. I hoped she'd made chocolate cake. There were only six hours of hard driving left. Darkness fell. Yawn, felt like the drone had invaded my blood stream. Then it happened: I fell asleep. At one hundred and twenty kilometres an hour!

I was startled awake as I passed under a highway lamp, mere inches from the outer edge of the shoulder. "*Tabarnak!*" I screeched to a halt, trembling and white as a ghost. Raging and fuming, I ripped off my pack and jumped into the ditch. In the blink of an eye, I'd thrown down my foam pad and bag. I slept, avoiding the confrontation with myself.

When the roar of a speeding tractor-trailer woke me, I fully realized how incredibly lucky I had been to have survived that close call. Fair warning. I will never forget where the true dangers in this world lie. I headed off to some nearby park to complete my night's dodo and imagined myself transposed into safe wilderness.

The next day I pulled into the familiar driveway in Scarborough, loudly honking my horn. Mom had no idea I purchased a motorbike. She barely peeked out the door, staring at the leather-clad, dirty biker with shoulder length hair and long bushy beard. She was obviously alarmed.

"Hi, Mom!"

"*Eh, quoi?* André, is that you? What the...? You know how to drive that thing? Isn't it dangerous?"

"Not at all, Mom!"

A few days later, after being coerced to trim up, I phoned up my long-time survival partner, hoping he was free the following week.

"Bert, can you extend your Thanksgiving vacation by a couple of days? Let's jump on the bike and spend a week out in the bush! Packing will be easy, we just won't bring anything at all!"

The plan sounded great to Bert, so when the weekend arrived we donned helmets and rode up to Dorset via Highway 35, stopped for fried chicken, then headed into Lake Kawagama territory. We followed progressively narrower backcountry roads until we encountered an ATV trail, into which we partially drove and partially pushed the bike. From the top of a hill we spotted an isolated lake in the distance, which suited our fancy. We shoved the bike into the woods and covered it with branches. At the base of a nearby tree, a hole was dug to stash keys, wallets, and the rest of our pocket gear. We hiked off to the lake, wearing nothing but the clothes on our backs, all pockets turned inside-out, for the time enjoying fine mild weather.

Our first task, fire building, was accomplished by rubbing sticks — after over three hours of intense work carving the bow, board, spindle, and handle by splitting and grinding with improvised rock tools. Good job! We were in cheerful ecstasy as always after accomplishing the miraculous feat. Next it was shelter time. Anticipating rain, we wished to strip nearby birch trees of their bark, long ago having learned that evergreen boughs won't shed any substantial downpour. Once hacked with sharpened stones, the bark from the small trees stubbornly refused to peel in wide sheets, since it split along the rough edges. Even a tiny metal blade would have been precious. It gave me the idea of trying to create one from the tab on my jeans zipper. I stripped to my underwear and

Banquet Time

managed to smash the tab between two rocks to remove and flatten it somewhat. Then I tied it to a wooden handle with a piece of shoelace and a constrictor clove hitch. After sharpening the tiny knife on a smooth rock, I was able to slice the birchbark from top to bottom and obtain the required shingles for a lean-to shelter roof. I marvelled at man's ingenuity, who through the ages learned to transform minerals into valuable metal. At the same time, I felt somewhat guilty to have "cheated."

By the time the shelter was completed and firewood gathered for the night, darkness had fallen. The only food I saw that day was a big cockroach-like bug that zipped out from under a log I had disturbed. It reminded me of how I once amazed my fellow prestidigitators at a convention when I held a deck of cards against a wall and the chosen card exited the deck, climbing the ten-metre high ballroom wall in a zigzag pattern. To this day, no one ever figured out how I accomplished that feat, probably because there was no secret — it was just luck. I had captured a huge cockroach in a damp corner of the washroom and had stuck a duplicate card to its back using magician's glue-wax. My lovely cockroach-assistant sure put on a good show for me that day. With those buggy thoughts I flopped onto my bough bed and philosophized for a long while with Bert, prevented from sleeping by pangs of hunger until late.

Growling bellies woke us early. A morning soggy with mist greeted us. The rain intensified but the birchbark roof mostly resisted its onslaught. We spent the day huddled in our cramped quarters, profiting from breaks in the rain to fix leaks and gather more wood. There would be no meal that day, but with the stars speckling the sky like dots on a trout, we looked forward to a good food-gathering the next day.

As the first rays of sun painted our surroundings golden, hunger drove us out of the shelter and the exploration began. A hearty feed of cattail roots would be most welcome, but after circumnavigating the lake from shore, there were none to be found. In fact we found no other edible plants either, except a couple of old bunchberry clusters hardly worth gathering. We plowed back through dense shrubs to the shelter and revived the fire, silent and psychologically distraught.

I headed out once more and aimlessly wandered for a while in the hope that manna would fall from the sky. And as I peered up it eventually did, in the form of a nice fat Spruce Grouse. It was hunting season, so I felt justified in harvesting it. What a treat it would be! Grouse are generally not suspicious, which facilitates their capture. Come on Bourbeau, don't miss your chance! Advancing slowly while keeping the future roast within my field of vision, I looked around for a long and slender pole. The live birch I chose to break seemed determined to remain attached to its trunk by wood filaments. I bent it over to the other side and twisted it with all my might, but having chosen it just a tad too big, the more I fought with it the more it resisted. I turned to a half-dead cherry branch that I barely managed to sheer off. Was the grouse still there? Yes, but it looked a bit nervous and was perched quite high.

The lace of my right boot would temporarily serve as snare wire. I fixed a noose to the end of the boom, but it kept closing itself or hanging limply. I finally tied the lasso a few centimetres from the end and tore a split into the tip of the pole with my teeth in order to wedge the lace loop open. That cherry bark tannin sure is bitter! Finally, I advanced with utmost caution toward the beautiful bird, undulating my body from side to side in rhythm with the wind. I did not approach it straight on, but rather at a forty-five degree angle until I was slightly past it, then repeated the process to get progressively closer.

Just a few metres separated us. I dared not move for a while. It was pecking at buds three metres above the ground. With great care, I raised my perch, a few centimetres at a time, nearer and nearer the soon-to-be barbecued chicken. Get the fire ready Bert! I elevated the tip of the pole. Just a tad more and the poultry would be mine. I managed to gently slip the collar around its neck and then purred to attract its attention, which made it raise its head to look around. I jerked down hard and quick. Damn! It flew away. How the hell did I miss it?

But it was still possible to get to it. I crept forward gently, as before. The bird was in view, but perched on a higher branch. On the tip of the toes, with arms outstretched, I tried to slide the noose over its head once more. But I couldn't control the shaky perch. The grouse thumped its wings and flew away beyond sight, leaving but the muffled echo of a tambourine. I can't say I was proud of myself as I returned to camp to cry on Bert's shoulder.

Banquet Time

Hunger hit hard. It wasn't at all painful, it just made us lazy, so very lazy. Our bodies reacted to the lack of nourishment by falling into survival mode, enticing inaction. Lousy sleep didn't help either. We napped and then lay there for quite a while afterwards before setting out in our quest for food again. That time an aquatic plant called Pickerel weed (*Pontederia cordata*) attracted my attention. There was plenty and I knew the seeds were edible. I removed my pants and socks, keeping shoes on to preserve my feet, and waded out in waist-deep water and muddy bottom to collect the plants. It was a waste of time and energy, as I soon realized the seeds were not at all ripe. I brought some back to try parching them on the fire, to no avail. Meanwhile, Bert had been weaving cedar bark into a fishing line and made a fish hook with his zipper tab, our new favourite tool material. He went off fishing while I tended to the fire to dry out my shoes. After an hour and some, he came home to the shelter empty-handed, as I expected.

Bert's fishing gave me an idea. I went around to the tip of the lake, where the water seemed deeper and rocks gave me shore access without wetting my feet again. There I spotted a few minnows. I sacrificed my t-shirt and spread it over a willow hoop to improvise a minnow net. But the cotton material was much too tightly woven and I couldn't pull the net through the water quickly enough to snatch any fish. Finally, through much patience, I managed to slowly raise the t-shirt from the bottom underneath the minnows to capture a couple. A dozen tries and nearly an hour later I obtained a grand total of five two-inch minnows, which I slit open and cleaned using my zipper-tab blade.

On the way back to camp, I managed to gather a half-dozen bunchberry clusters and a few violet leaves. As I let the small fry cook on a hot rock, I prepared two birchbark "plates," two sets of automatic chopsticks, and then served up our minuscule supper of two and a half minnows each, delicately decorated with the violet leaves and berries. Banquet time. Bon appétit, Bert. Another occasion to philosophize with my good buddy!

The next day we used Bert's fishing line to set a few rabbit snares, with no high hopes of success, considering few signs of their presence. We scavenged some more, but as the shadows stretched into long thin lines, it looked like we wouldn't be ingesting any more calories that day either.

After another uneventful but chilly night of alternating periods of sleep-wake, I decided to head further away from the lake, back near the road, hoping to find food in the field-like environment found there. Indeed, I did find a few leftover raspberries spread out here and there near the roadside. After an hour, however, no more than a cup of the precious fare lined the bottom of my hat, and I wondered whether or not the calories they contained would outweigh my energy expenditure. I was slouched, pondering, intensely focused on my berry picking, when out of nowhere a loud grumbling startled me. A car! Instinctively, like a wild animal, I dove deep under the bushes to hide as the roar passed by and subsided, leaving a cloud of gravel smoke to fall upon me. I recovered from my emotion, laughing deeply at my new understanding of why we never see fauna while driving.

Returning to camp I shared my few berries and story with Bert, replacing him at fire watch so he could go forage too, effectively inverting our picket and picker roles. My weakness forced me to nap. Later, Bert woke me to partake in another half-cup of delicious berries. We spent the rest of the day exploring further still, trying every trick in the book to obtain food, but nature simply wouldn't cooperate. The next three lethargic days just saw us get progressively weaker. We were no longer having fun. But it was our last night.

The sun lighting up the morning dew transformed the water drops into jewels as if to thank us for our visit and wish us well as we departed. Gleefully heading back to colonized land where fridges bulge with offerings, we concluded that wilderness survival is largely a matter of luck.

Batman Again

"You have to take risks. We will only understand the miracle of life fully when we allow the unexpected to happen."
— Paulo Coelho

In my early twenties, I tried to gain new outdoor skills by simply multiplying time in the field. But I noticed I was mostly gaining experience through all the slip-ups I had to rescue myself from. "Live and learn," as the old saying goes. I definitely learned about prevention the time I paddled the entire Highland Creek from Highway 401 all the way to Lake Ontario in a minuscule toy rubber dinghy — during spring flood. That was a wild ride to be sure, especially the last half with my finger plugging up the hole created by smashing into a floating hawthorn tree. One needs a patch in those situations.

Long before the advent of modern mountain bikes, my best buddy and I built ourselves some trail bikes using old newspaper delivery bicycles, to which we had welded the low gearing from some children bicycles. We covered a lot of ground on those contraptions and were many times stuck out in the middle of nowhere with flat tires. I wonder today why I didn't carry a patch in those situations either; I no doubt enjoyed the challenge of being in a jam out there.

1973. The first mountain bike?

It was when I was standing up in a toboggan zipping down a steep ravine that I learned that bones do break. And skis break too, as I found out on the sixth day of a winter camping trip around Algonquin Park's longest hiking trail. With four days to go, I managed to switch the binding to the front half of the broken ski, but when the half-ski broke in the middle again, the only way to continue was to build a primitive homemade replacement from a maple branch. A patch doesn't help here; one needs a full-fledged repair kit.

I suppose it's only after flipping the canoe in rapids and losing stuff downriver a few times that one learns to tie in the gear properly. At least that's how it worked with thick-headed me.

I'll stop here, simply to maintain my shame at an acceptable level. There were countless mistakes. Chalk up one thousand to experience!

Okay, one thousand and one. At Taylor Statten Camps, I had been asked to start a bow-drill fire in front of the council ring of over three hundred people. To impress the gallery, I had the brilliant idea of pre-filling the notch in the board with burnt wood powder, figuring I would then only have to spin the drill enough to create an ember; I had glorious visions of fame from my would-be record-breaking speed. To make a long story short, the effect was opposite. The cold wood powder completely prevented the formation of the ember, so I was left there without a fire and without a place to hide my head in shame. This humbling experience really ticked me off, and for the rest of the summer, fuelled by that powerful substance called ego, I practised fire by friction with every possible variant — all species of hardwood, left handed, even blindfolded — until my fingers holding the bow string taut had developed thick calluses. On a positive note, at least the knowledge and skills gained by all of these blunders became impregnated in me, as I paddled, hiked, climbed, skied, snowshoed, biked, and camped my way to becoming an outdoor professional.

In my mid-twenties, during my graduate school years, I also sojourned in Third World countries to broaden my survival skills horizon. My first major trip was to Africa's Niger in the Sahara desert, then the second poorest country in the world, where I spent nearly two months with the Hausa and Touareg people. A food and outdoors background had landed me a job in an international cooperation project with the responsibility of documenting local food habits and survival techniques. During this life-changing experience, not only did I acquire specific skills like starting a fire with millet stalks with the hand-drill technique, or weaving mats and hats from palm leaves, but I also experienced a major sand storm and was able to join a camel expedition. Other travels included adventures such as hiking across Haiti and witnessing mind-bending voodoo ceremonies, hiking to a remote mountain village in India with a local doctor to spend a few days with a tribe that had never laid eyes on a white man, or going deep-sea fishing in the Bay of Bengal on a wooden sailing raft built from logs. What blew my mind during that experience was that these fishermen would actually wedge a perpendicular plank between the logs so they could trapeze while sailing.

Another memorable adventure was descending Costa Rica's Tortuguero River via dugout canoe in cozy proximity to crocodiles. A highlight of that trip for me was one night when the local guide jokingly dared me to catch one of the reptiles as we spotted the red glow of their eyes with our lamps. In a flash I had pulled a snare out of my pocket, tied it to my paddle, and hoisted a four-footer into the pirogue. Somehow the noose had only

Leaving for deep-sea fishing and sailing on the Bay of Bengal.

hooked it by the snout, and we had quite a time trying to return it to the water as it trashed about and snapped at our feet.

By the time I had completed my doctoral degree course work, my curriculum vitae had also been enhanced by a third year in a high school classroom, this time as a replacement science teacher. In reality, all I remember of that stint was getting the students on my side by letting them experiment with distilling moonshine, but it looked good on paper. My employment as a nature guide receiving groups of school kids at the Claremont Conservation Area also bolstered my teaching resume. Because of all this and flattering letters of recommendation I had obtained from former employers, project supervisors and graduate school staff, at the age of twenty-eight I was offered a job as professor of outdoor pursuits at the University of Quebec's Chicoutimi campus. Nice way to pay off my student loans, I thought. At the time, little did I suspect I would spend my entire career there.

I admit I was just as delinquent a young professor as I had been a secondary school teacher. I recall one incident at Thornlea High School, where during a karate demonstration in the middle of math class I had inadvertently kicked a hole right through the plaster board wall, sending the entire bookshelf in the next classroom crashing down and just missing the stuffy old history teacher. "A demonstration in leverage gone wrong, so sorry," I justified. And sorry I was too at my first university department meeting, when a roomful of straight-faced faculty members witnessed me fall backwards over my padded armchair as I exaggeratedly fidgeted and rocked from boredom.

Most students loved me though, especially the gifted, who appreciated my unique way of questioning the status quo. They would repeat my favourite advice: "Don't let university interfere with your education!" And most of all, we had f-u-n, in and out of class. For example, every day they would hold the doors open and dare me to break my own record as I donned my mountain bike and helmet and rode from my office, down four sets of stairs, and then straight outside. In class they appreciated my trust as I would give them all A's from the outset and then ask them to earn it. This actually worked; too bad the dean didn't agree with my unusual methods and forced me to change.

The first couple of years at the university were spent struggling to complete my doctoral dissertation. Once that was out of the way, I felt it was time for creative pedagogy.

"What do you mean I can't teach my outdoor leadership course in Mexico? It's too late, all the students have already financed and bought their tickets and we're leaving next week on our caving expedition!"

"The University's insurance program doesn't cover out-of-country activities, and besides, you didn't advise me of this project ahead of time."

"I didn't realize I had to."

"Hey, I'm the department head!"

"Okay, I guess I should have informed you of this. Didn't know. What do we do now? The students have worked for the last three months preparing for this course."

"Well, I simply cannot authorize this departure. Sorry."

I left the stark-faced encounter pondering the value of my delinquent ways. Damned insurance! If it weren't for this detail, the department head would never have found out about the trip and we would have been free to leave. I couldn't cancel; the students would kill me. Three full months of planning and research. I was sure I had thought of everything. There hadn't been a word about insurance in Colorado when our botany professor took us on a field trip. Why was I in this mess now? I barged into the dean's office, while calling out to his secretary: "Emergency!"

The dean heard me out, but couldn't or wouldn't do anything for me. Stupid politics! Off I went to the vice-rector's office. He was in a meeting somewhere. I convinced his secretary to squeeze me into his schedule the next morning. I went home to ponder. I pulled out the dossier I supervised, where the students had laid out in great detail the logistics and the security procedures. I read it one last time, satisfied. The next morning I met the vice-rector, dressed in my only suit — my politician costume — and charmed him into submission by emphasizing the fact that our role is to support the students in their endeavours and citing examples from other institutions. Being open-minded, he immediately acknowledged that I'd done my homework and thus agreed to take it upon himself to personally support the trip and to make the arrangements for insurance. I left in disbelief, relieved.

A week later we landed in Cancun, a world-renowned tourist trap. I knew we had to get out of there quickly, or my students would lose the feeling of disorientation I was trying to create. We immediately squeezed onto a local bus headed for Puerto Juarez, a small native village away from tourists. I could tell by the eleven students' faces that they were overwhelmed by what they saw. I too was weighed down by the trip. It was only my seventh time in Mexico, *mi Espagnol no es bueno*, and I felt extra pressure to make this adventure successful. I booked us into some local hotel for less than ten bucks per room. After a supper of tortillas and bananas, safe food, we crashed for the night, exhausted by the journey. The next day, under student leadership this time, we bussed to the town of Mérida, where we visited the university and museum. Then we bought hammocks and camping food for a few days. Secretly I bought a devil's mask I spotted in a stand; it would come in handy down deep in the caves.

When the roosters woke us the next *mañana*, we visited the archaeological ruins of Uxmal and then took a stuffed chicken bus to Villahermosa. There the students hired a local guide to walk with us to the entrance of the *Cuevas de Xtacunbilxunan*, near Bolonchenticul de Rejon, where we camped by stringing up our brand new hammocks between thorny trees. We also arranged for the guide to cook up a whole piglet barbeque for our supper. We partied *sin cerveza*.

The strident yell of monkeys woke us just as the sun illuminated the incredible entrance to the wild caves. I pulled out the old caving guidebook unearthed through inter-library loan. It appeared no one had visited the particular cave we wished to explore since 1973. The main attraction was the underwater lake at the end of a two-kilometre long passageway. I picked that particular cave because it harbours but a single passageway, so there was a reduced chance of getting lost. The lake sounded neat, and most of the students were quite excited at the prospect of swimming in an underwater cenote. Eight decided to partake in the adventure. The other three — more tourists than too-risks — would watch our campsite. We tossed the packs brimming with climbing gear and lamps onto our backs and headed for the entrance, which came into view exactly where the guidebook said it would. So far so good.

Our first obstacle was to rappel down into the dark abyss. I went first with a powerful headlamp and reached the end of the rope before I

The entrance to the Xtacunbilxunan caves.

touched the ground. I tied myself to a prusik knot and then looked down. False alarm, the ground was only a half-metre below me. I jumped, then slid five or six metres down a scree slope to a flat area. I whistled up for the next person to descend.

The eight students with me were all adventurers in their own right. They could take the pressure. So I decided to have a bit of fun at their expense. When the next student reached the end of the rope, I was way below him. I yelled at him insistently to jump, which of course he wouldn't do since he thought it was a long drop. After letting him hang uncertain for a while, I climbed up and touched his feet, telling him it was a joke. He hopped down with relief. Then there were two of us to tease the next guy, and the last person was greeted at the end of the rope by eight loud summons to jump down into the non-existing void.

The student leaders that day had asked me to act as front-runner. We entered the creepy hallway and progressed through several chambers until we were back to a narrow passageway. At its end, another stalactite/stalagmite filled room, and another beyond. Several bats hovered in both. I was Batman again. The students took a pause. I told them I'd go ahead to answer nature's call. I didn't go back. Instead I put on the devil's mask and hid behind a low rock with just my head sticking out. As the student leaders came forward calling my name, I flashed my light onto my face. More fun.

Finally we got to the entrance of the narrow artery that lead to the lake. The passage reminded me of the squeeze hole I had experienced in the Missouri cave, but not nearly as tight. It was still quite low and we had to crawl on hands and knees to get by. After twenty minutes, the passage swooped lower so we had to pull ourselves along with nothing but our elbows. At least the ground was soft and mushy. Time passed. Still no sign of the lake. I forged ahead, followed by my brave students. We crept along until finally feeling the humidity and coolness of the approaching lake. Then we heard it. An eerie shrieking sound.

It was bats — thousands and thousands of them. All gathered at the underground lake, probably their only water source. As we approached they got excited and moved in huge swarms. It was one of the most powerful displays of nature I had ever witnessed. We all listened in awe, but

Batman Again

The passageway to the underground lake.

the students expressed concern. I shoved myself forward some more, sensing the lake was just there, a couple dozen metres away. The bats were more and more aroused, as they obviously could detect our presence. Some bats were coming toward us. It seemed like the path we were going through was their only way out. And all of a sudden, they ALL wanted out. A continuous flow of bats filled our narrow passageway, striking our hardhats and our shoulders as they flew by.

"There's one in my shirt!"

"Me too!".

"Just keep your heads down, guys, it can't last long!" I replied.

After a few long minutes it was over. We walked to the edge of the lake, which was nothing but a mucky, dirty hole. The stench was unbearable.

"Let's get out of here!"

The crawl back out suddenly became painfully disgusting, as we realized the reason for the soft ground below us. Guano! A lush carpet of bat excrement!

Needless to say, only fresh air and a good cleanup could transform us into happy campers again. Happy campers with new stories to share.

The rest of the trip remained rather uneventful, probably because we didn't see any jaguars when we hired a guide to take us in a motorized dugout to "hunt for tigers" in the jungle near the Belize border. We finished our Yucatán tour with a bit of wonderful snorkelling on the barrier reef, then too soon we were back in Canada.

A few days later, I was soaking my sheets with fever. Never had I been so sick. But after a couple of hours had passed I was just fine and commuted to the university. By lunch I was back in bed, with incredible fever and headaches again. At suppertime I felt okay. Then came night chills followed by more menacingly high temperatures on the thermometer. Weird, powerful flu-like symptoms appeared and disappeared suddenly. The doctor brushed it off. After all, I looked perfectly well when I went to the clinic. In another hour, I was totally wiped out again.

When I went back to the university the next day, between two spells of fever, I met a couple of my students who reported the same symptoms. A few phone calls later, I deduced that only the eight students that had been in the cave were affected. I ran to the library and pulled some medical books off the shelf. Looked up bat diseases. There were two. Histoplasmosis, sometimes deadly in old or weak persons, and coccidioidomycosis, often deadly. Oh my God! I jumped in the car and headed back to the clinic. Without an appointment, I saw a different doctor. This one witnessed me in fever mode and I asked him to test me for histoplasmosis and coccidioidomycosis. The x-ray showed completely white lungs. He was most impressed, showing it to the other doctors. None of them had seen this before. Scary moment.

Turns out we were affected by severe histoplasmosis. Histoplasma fungus lives on bat guano and we had breathed in the spores, which later grew in our lungs. We were ill for over two weeks before the symptoms subsided. The slight consolation was that apparently we were immunized for life, although benign white scars would forever remain on our chest x-rays. Makes for good conversation with medical staff.

I never filed the final report for this trip. Blamed it on the flu.

Fifteen

A Cold Hole

> "The cure for our modern maladies is dirt under the fingernails and the feel of thick grass between the toes. The cure for our listlessness is to be out within the invigorating wind."
> — Lucy Maud Browning

The fungal souvenir brought back from Mexico's caves gave me a brand new perspective on life, which generated ideas for other types of survival adventures. Thus far I had always enjoyed perfect health during my experiments. But what if? What would happen if I were sick or injured during a survival ordeal? Scary. Ah, but it is only the unknown we fear. I figured that to remove my apprehension, there was but one solution: make the unknown known. The time had come to stare illness in the face, to test my guts, both literally and figuratively.

The following October, the opportunity proudly presented itself as I was suffering gastrointestinal grumblings accompanied by a high fever and general stuffiness. Lying in bed miserably sweating that night, I reluctantly recalled my promise to myself. An hour later, I was stumbling about a dark forest in the Laurentide Wildlife Reserve desperately attempting to light up my damp and gloomy surroundings by strategically starting bark fires on top of old stumps. Gladly I was blessed by a warm spell and within a couple of hours, despite the splitting headache, I had gathered a substantial pile of firewood and lay slouched on a bed of fragrant fir boughs besides a familiar cheery fire. The exercise-while-sick effort had wrung me out like a wet towel, and I flopped limply into a deep sleep. During the night, the humid fresh air worked wonders on my stuffy sinuses and as I briefly half-woke to rekindle the fire I recall feeling that my germs were disappearing into the wilderness rather than being recirculated in an enclosed bedroom to attack me over and over again. I slumbered through the entire next day and night, exiting my bed only to slurp a drink from the nearby stream, and woke the third morning like a new man.

This success encouraged me to try other similar challenges, imagining various handicaps for myself. One night I would enter the wilderness with crutches as if I had a broken foot and when I survived that easily I would up the ante on the next experiment by tying one end of a twenty-metre-long rope around my feet and the other end to a tree — to simulate having to survive by crawling with a broken leg — all the while considering myself lucky to be pain free. Another time I dared myself to start a friction fire with one hand tied behind my back, using my teeth to hold the drill handle steady. On still another occasion, the test was to light the fire using a book of paper matches and just my toes, a quite frustrating endeavour to be honest. But certainly the most out-of-the-ordinary experiment was surviving twenty-four hours blindfolded on a freezing night dressed in light clothing; that sure made me appreciate the oh-so-wonderful gift of sight!

These what-if scenarios inspired me to concoct others. I convinced a group of my students to try surviving a weekend inside an area delimited by a climbing rope tied in a circle; that ended up being a great exercise in leadership and also in environmental awareness, as we stripped the area bare in no time. Pretending to fall through the ice by jumping into frozen water fully dressed and just before dark also made for a worthy challenge, as did surviving a night while remaining enclosed within an alder swamp with water up to my knees. I don't know exactly what my motivations were. Probably a combination of the performance monkey still looking for the peanut, a true desire for challenge, a profound love of nature, desire for freedom from society standards and a search to become even more like Jules Verne's Cyril Smith.

A Cold Hole

Sometimes, though, I gained even more through a bit of friendly competition....

"Hey Mario, what do you think would be the better tool to survive out here tonight if we only have the clothes on our back and our pocket gear? Would you rather grab an axe or a shovel?"

"A shovel, no doubt about it!"

"No way, I'd much rather have an axe. That way I can build a fire. Besides, I can shovel with my snowshoes."

"Nah, the snow is way too deep and the firewood too scarce. I'd rather build a snow cave. Need a shovel for that!"

"I still would rather take my chance with a fire.... Betcha I would suffer less than you in a snow cave!"

"No chance!"

"Put your money where your mouth is. Let's do it!"

My great buddy Mario happened to be my partner and fellow outdoor professor at the university. He's always up for a challenge, especially when it relates to practising in the cold. Mario is a passionate alpinist, having climbed the highest peak of all seven continents and looking for a second chance at the summit of Everest, which he barely missed last time due to the heroic deed of saving someone's life instead. He knows his stuff.

The challenge we were talking about seemed unreasonable, even for Mario and me. It was minus forty degrees — excruciatingly cold. It was the last day of the winter camping course we were team teaching, and as soon as the chili supper was gulped down that evening, all the students vanished into their Arctic tents to seek the refuge of their thick sleeping bags. Mario and I were standing in the dug-out snow kitchen putting away the cooking paraphernalia. The full moon outside was casting eerie shadows of the snow-covered tree phantoms, many of which were emitting frost-caused groaning noises. The orb's reflections on the earth's white mantle illuminated the mummified forest as if in broad daylight.

This well-known valley tucked away behind the Valin Mountains in the Saguenay region receives record amounts of the white stuff. The previous year, the weather station's five-metre-long measuring stick disappeared from sight. It was just mid-January and already we were surrounded by three metres of powder. Mario was right, the wood was indeed scarce, and buried deep; perhaps he was also right that a snow cave is the better option. We would soon find out, as we headed nearby in opposite directions, after having made a last round of the student tents.

I saw Mario voluntarily stepping near a short spruce piercing the snow and falling deep into the tree well, something we usually meticulously avoid. Snow flew out in a spray as he started shovelling his snow cave. His slow rhythm was one of experience; he certainly didn't want to work himself into a sweat, his dry clothing being his only asset.

On the other hand, I didn't have to be so careful since I would have the opportunity to later dry out by the blaze. My first task was to dig a ten-metre-long trench right down to the ground. To ease this job, I found three spruce trees lined up next to each other and enlarged the tree wells at their base by shovelling with my snowshoes. Next I downed a convenient but half rotten spruce tree and chopped the sound sections of it into slivers. I managed to start the fire in the middle of the trench on a bed consisting of the remaining shattered pieces.

Next I scrounged for wood, which I had to gather from a large area. To save energy I used the Timberjack technique. This consists of creating one long and straight main trail that soon becomes a hardened path. I then toss nearby dead logs into this main trail where I can lug them back to camp using a harness made by tying cords to each end of my belt. Thusly I dragged a half dozen of the longest pieces to the trench and placed the ends above the ruby coals. At first the trench filled with overflowing smoke, forcing a coughing exit, but I was convinced it would clear as the logs dried out. Meanwhile I retraced my steps to the very end of my trail and stretched it out farther to fetch more and more wood.

When done, I entered my trench and tended the fire to warm up my cold paws. I noticed Mario busily breaking spruce branches here and there and transporting huge piles back to his chosen spot. That meant he had completed his snow shelter and was getting ready to line its floor with boughs. It looked like he was managing fine.

A Cold Hole

I moved on over to spruce trees for boughs, carefully avoiding inadvertently falling into the trap of a tree pit. The axe facilitated my job and I was done at about the same time as Mario. Game even. Both of us lined our shelters with boughs. I simply enlarged the space melted away by the heat of my blaze, using the wet snow to create a wall behind me. Like Mario no doubt, I was sitting on my snowshoes on top of the cold evergreen branches. Unlike Mario, however, my mitts were soaked and I was shivering with an undershirt damp from sweat. Ah, but I have the advantage of fire, which permitted me to slowly dry out the mitts and the front side of my base layer by unzipping my nylon overcoat and unbuttoning my thick wool jacket. After a while I slipped my arms out from under my clothing and rotated the shirt at the neck to heat its back in the same way. Finally I was dry.

But not comfy. It was just too chilly on my backside, and the spruce wood tossed a million violent sparks my way, forcing a continual squinted watch. I wondered how Mario was faring in his tiny cramped and pitch-black quarters. From experience, I knew that he could hardly sleep either, because his body heat would only raise the temperature in his shelter to within ten degrees of freezing. He would have to perform calisthenics on and off all night long. I thought I preferred my own fate, as I slouched down for a short-lived nap.

Several hours of torture went by. Three o'clock. Damn it was cold! I just couldn't let up on tending the fire at all, or I would start to shiver and shake. Then out came Mario.

"Nice fire. How are you doing?"

"Pretty good. How about yourself? What are you doing up and about?"

"Had to get up to take a leak. It's a lot warmer inside my shelter than out here that's for sure!"

"Come and sit down by the fire for a while, it'll warm you up."

"Nah, no thanks. I'm fine. I'll just go back to my warm cave."

"Okay. Suit yourself my friend, but it sure is cozy here by the fire."

As Mario returned to his cold hole, we both knew only too well that we were lying through our teeth. I truly love Mario, he's like a brother to me. I was so glad to share this adventure with him; we would no doubt reminisce about it for years to come, teasing each other, each of us insisting we won the game.

An hour later, after I was sure I did gather enough wood to make it through the night, I cranked the fire up a notch. In the fetal position I turned my cold back to the flame as increasing fatigue finally pulled my eyes shut. Three more fire-poking stints later the sun finally decided to peek over the horizon and show its bright warm face. After a while I watched as Mario joined the most eager students who were already up and about chugging cowboy coffee and cooking a couscous breakfast. Since Mario was supervising the students, I slumbered a bit more while the crowd gathered under the outdoor kitchen tarp. Finally it was my turn and I appeared in the pit entrance, still feeling the chill on my backside. Looking at Mario's bleary and sunken eyes, I was convinced I'd prevailed this time as we entertained the students with each of our stories, while also instructing them on survival strategies in case of gear loss.

After our speech and instructions for the return trip, I turned to go pack up my own unused tent and sleeping gear. Just then the crowd roared with laughter.

"What's so funny?"

"It's your jacket! Check your back!"

I pulled off my overcoat and stared with disbelief at the gaping hole burned right through the nylon fabric from shoulder to shoulder and neck to waist. No wonder I felt a chill. A cold hole, colder than Mario's. He won!

Sixteen

Ferns Crack Me Up

> *"So many tangles in life are ultimately hopeless that we have no appropriate sword other than laughter."*
> — Gordon William Allport

It seems to me that psychology professor Allport is another person I have proven right on too many occasions. Sometimes outdoor life presents us with such strange and unexpected situations that we just have to take them in stride and laugh. Once while simply canoeing along with my family on Ontario's historic French River, we had to dive for cover between two short cliffs when a wild and whirling storm appeared out of the blue. We landed in panic mode and I quickly dragged the canoe onto shore and tied it to a rock. As the drops began falling on the violently shaking trees (and the violently shaking us), the only spot we could pitch the tent was right on top of a dense patch of poison ivy. Laughter took over as we tried to guess who would start scratching first. Trees toppled nearby while we supported the tent poles with our hands to prevent their breaking, and it seemed like the fabric protecting us would be torn to shreds. At the end of the trip, news of tornado damage informed us of our close-call. As for the poison ivy, we had wiped ourselves and our tent with Jewelweed juice, *Impatiens biflora*, a well-known natural prophylactic and cure for this ailment. We were fine.

While I was with my lady partner and sled dogs on the fourth day of a winter camping expedition, unusually hot weather so softened the deep snow that each snowshoe step felt like hoisting oneself onto a table, only to sink back down again. We managed to advance no more than half a kilometre that morning, and when forced to camp by mid-afternoon, chores such as tying the dogs to their chains or setting up the prospector's tent transformed us into the sluggish actors of a laborious slow-motion film. The warm spell lasted six days and we were unable to proceed. We laughed and joked that we would still be there the next summer. When a cold spell finally liberated us from our soggy prison, we literally flew home behind super-excited dogs, enjoying a wild ride over icy hard surface.

Another time a teaching assistant and I were making an advance visit to the Laurentides Wildlife Reserve for a forthcoming winter-camping expedition with students. The blowing snow became so deep the dogs couldn't break trail, so we alternated in front. After having spotted our campsite locations we retraced our steps, only to find the trail had been mostly refilled with powder in such a way that we snowshoed with one foot high and one low — backbreaking work. At minus thirty-five degrees we pushed on through the blizzard, letting ourselves get sweaty and exhausted with the knowledge we would soon toss the gear and the dogs into the truck and trailer I had parked on the shoulder of the highway. But when we arrived past midnight, there was no warm haven to be found. The police had towed our getaway vehicle! Bastards! We tried hitchhiking for a few minutes, but the occasional tractor-trailer trucks speeding by would only see us at the last minute and ignore our presence. We were freezing fast. And even if someone did stop, I couldn't exactly slip the dogs into my backpack.

We had a good laugh as we overturned the sled to block the wind and stripped to bare torsos to quickly change into dry clothes. That was quite the polar-bear initiation! After performing dozens of jumping-jacks to regain heat, we clambered aboard the sled runners and I yelled to my lead dog: "*Marche*, Kilou!" Off we went down the highway and back toward town, some twenty odd kilometres away. Again we laughed when the first van passed us by, imagining the driver's face as he encountered our unidentified sliding object. In fact, we giggled until we gained the edge of town. It was four a.m. When low and behold, there was my van and trailer locked up behind the gate of a towing firm's lot. I picked the lock and took off without further ado. The next morning I phoned the police and criticized them severely for their action, raving and ranting that a less experienced person in that situation would have surely died. "Give me a stiff fine," I said, "but don't take away the vehicle I depend on for my safety!" As a reply, I was informed that my licence payment was overdue and the fine for driving without a valid document was three hundred and fifty bucks. Oh. Finally, the officer felt guilty enough to let me get away scot-free. I never heard back from the towing firm, either.

Other real-life mishaps seem so very serious when they happen, but give one great comic relief once safe and dry. Like the spring I wrapped my fibreglass canoe around a rock and watched helplessly as it was ripped to pieces by boiling current in the raging, flooded St-Jean River. I ended up on the snowy shore dressed in nothing but a wetsuit and only an emergency flint striker hanging around my neck. It was a long painful bushwhack back to the car. Or another similar incident, again during a pesky spring flood, this time on the Metabetchouan River when my canoe was engulfed in a hole the size of a bus. Climbing out of the canyon to go rescue my craft somewhere downstream, I grabbed at a boulder-sized rock with my right hand, which then unapologetically detached itself from the cliff and crashed onto the rocky shore below. I was left hanging from a root, holding on for dear life with my left hand, like something out of a movie minus the off-camera safety net. A very terrifying close call, but a great anecdote at parties, especially when acted out. I never did recover that canoe.

Another heart-pounding encounter with death occurred on the Mars River, after I had dropped my rope while lining a serious class 4 rapid. To reach the canoe wedged on a rock just above a two-metre drop, I had to fell a huge spruce tree and was quite pleased when it landed directly on said rock. I peeled off my wetsuit down to the waist, as my body had heated up from chopping, and unwisely simply tossed my lifejacket over my shoulders to cross over to my canoe on the trunk bridge. As I neared my goal, my weight bent the tip of the tree. When the branches touched the water I was swept away with the tree and down the chute. I was unhurt, but had lost my glasses. The freezing water filled my wetsuit and was dragging me down, especially since my flotation device was unzipped. To make matters worse, a ten-metre waterfall waited right around the corner for a victim. I reached shore just in time, petrified. Many years went by before I could laugh at that one.

Another incident that took a while to digest occurred on the Saguenay River, during a simple ramble to paddle around pretty floating ice blocks on a late sunny April late afternoon. I was teaching a friend how to paddle, and I admit, trying to impress her with my powerful bracing strokes. In a moment of exaggeration, with her sitting up on the bow seat, the paddle cracked and in we went. We were mere metres from shore and dressed in wet suits, so we laughed at my "bad stroke" of luck. But the wind pushed us away from the point of land nearby and down into the bay. No problem, we would just float down to the shore on the other side. This took longer than expected, and cold was slowly gaining on us. And then, suddenly, it was dark. In the far distance, spots of light dotted the horizon, and I started to wonder if the wind had shifted and was pushing us beyond the bay. Forty-five minutes later we became more and more hypothermic, the heat leaving our bodies. Just as I thought we were done for and considered using my last remaining strength to tie our wrists together across the canoe to extend our survival time, my feet touched bottom. I flipped that canoe over as if possessed by superhuman strength and we flopped in to start paddling to shore, laughing. Just then my friend started convulsing in the bottom of the canoe and screaming bloody murder. I tried to comfort her with useless words: "Don't worry, stop yelling, we're safe now!" But a minute later I was the one yelping in pain; our entire bodies were thawing out, and we were experiencing the pins-and-needles of the process. A warm, fully clothed shower in a nearby house put us out of our misery.

I couldn't count the number of times I had felt the warning sign of those pins-and-needles on my feet. Every time I would smile knowingly: "Here I go again, playing a game of lose-your-feet-or-not with Mr. Frostbite!" The most painful occasion was during a ten-day backcountry ski trip across the Grands Jardins Park. We woke up one morning to minus thirty-eight degree cold and my leather boots were stiff as rocks, in spite of having placed them between two sleeping bags overnight. As usual, we all wore our boots inside our parkas next to our stomachs during breakfast, and when all were just about ready, we yelled "Boots on!" to signal our departure together. I was responsible for folding the group tarp and started slightly behind the others, having a helluva time slipping my feet into my hardened footwear. Then I skied as fast as I could in an attempt to increase circulation, but to no avail. Mr. Frostbite was winning the game for the first time, as I was moving beyond the tell-tale pain into the frightening doesn't-hurt-anymore zone. With no one around to lend a warm stomach to lean my feet against, I seemed out of options. Then a benevolent birch tree appeared as saviour, with its miraculous bark. I tossed a huge pile into a heap and lit it. Sitting on my pack, I took off one boot and warmed my sock, then replaced it in the footwear. I repeated this with the other foot. Repeated again. And again. At last the pins-and-needles reappeared, accompanied by their excruciating but welcomed pain. After a few short minutes the bark was consumed, but I had won the battle and skied off to join my worried buddies.

The ways to get into trouble in the wilderness are countless. Never could we imagine all that can go wrong. It's the famous Murphy's Law and its corollaries: "Anything that can go wrong will, when least expected and at the worst possible moment." You'll get a flat tire when dressed in a tuxedo, in the rain, and late for your own wedding. Then there's Bourbeau's Law: "Murphy was an optimist!" When you get the flat, there will be no taxis or other cars around, and your cell phone battery will have gone dead. The jack will be missing from the trunk. And you will have locked your keys in the car.

On another occasion, I had just pulled what at the time I considered a great joke on my students. They were canoeing behind me, following me blindly during a leadership course. I had made them portage over an island, which they only realized it after I suggested to look back at where they had come from. So that evening I was half-expecting a practical joke as revenge. Then it hit me. I was sick. Oh, so indescribably sick. All night long I was bent in half with gastrointestinal cramps. Didn't sleep a wink. Details unnecessary.

The next morning I could not face my class. No one could have done this to me voluntarily. My assistant took over as I lay motionless in the bottom of the canoe. That night I got better, and the next morning I was fine. Some kind of food poisoning no doubt, but I just couldn't figure it out; I had eaten the same thing as everyone else, and we were all drinking pure river water. The answer revealed itself the following week as I was cleaning my gear room and happened to fall upon the wrapper of the new sponge I had shoved into my bailing bucket. The wrapper read "New! Pre-soaped sponge." I had taken a drink from my bailing bucket, without removing the sponge. A swig of concentrated soap. Not revenge — karma.

Outdoor life so often points out our shortcomings, sometimes I swear Mother Nature has a mind of her own and gloats while teaching lessons. I have always loved designing my own gear or modifying that which already exists. Perhaps I coast on the slippery illusion that I will invent something important — wishful thinking at best, but you never know. One day, I had reasoned that there was no need to carry both a sleeping bag and a down parka with pants. In an effort to reduce weight I modified my down mummy bag by sewing in closable slits for arms and feet. I figured that this "improvement" would permit me to just get up out of bed and have breakfast with my bag still on, leaving me nice and warm. I tried out my idea during a winter camping expedition in the month of March. As luck would have it, the days were warm and the nights cold, leaving us with a solid snow surface to walk on in the morning. I exited the tent that morning dressed in my "outfit," to the great amusement of my buddies, especially as I walked toward them penguin style. After a cup of hot chocolate, I waddled toward the designated pee spot, which happened to be at the edge of a hill. That's where I tripped of course, and I slid down the steep incline at neck-breaking speed on my nylon shell, my penguin feet and arms flapping wildly as I tried to slow the descent. Luckily, I landed safely in a bushy fir tree, to the thunderous laughter of the audience. I deserved that one, I suppose.

Sometimes in nature, the most hilarious moments occur spontaneously, without need to have been provoked. Especially when my brother Michel is involved.

Ferns Crack Me Up

The phone rang in the GB Catering laboratory where I was busily inventing and testing tripping recipes. It was my brother Michel, and he was in panic mode.

"André come quick. My pigs have escaped!"

"What pigs? What the hell are you talking about?"

"I bought these four pigs a couple of weeks ago, I'm feeding them with warehouse leftovers. I built them a pen but now they've escaped into the forest! You have to come and help me find them!"

"Sounds like fun to me. I'll be right over."

Michel lived in an old run-down cottage on a piece of land south of Huntsville in Ontario. Twenty minutes later, I entered. In the one-room cabin there stood a loud fridge with six cases of beer leaning drunkenly on it. A beat-up couch slouched against one wall, and a bunk bed with tons of junk on top hid the other. Milk crate shelving acted as furniture. Oh yes, and a bird cage with some bats in it decorated the corner. But what really stood out was the brand new pool table smack in the centre. A real bachelor.

"Hey Ho! Michel, where are you?"

"Over this way. I'm at the pig pen."

"Coming."

Out the back door, I passed an abandoned tractor, a few car carcasses, and a huge pile of greyish wood boards, remnants of a fallen shed. At the newer shack beyond I found Michel with two of my other brothers, ranting about the intelligence of swine.

"They're smarter than you, that's for sure. They got away!"

We all laughed.

"I don't believe it. I thought this pen was bombproof."

"Now what?"

"We have to wait for my two buddies to get here, then we'll go looking for them."

A car screeched to a halt. Michel's two scruffy buddies jumped over the doors of their convertible and joined us.

"Hi guys. What's up?"

"We have four pigs involved in this drama here, all of which are more intelligent than Michel!"

We were cramped with laughter.

"Let's grab a beer!"

"No no, this is serious guys. We've got to find those pigs. They're expensive you know."

I explained common search and rescue procedures to orient the men the best I could. The plan was to first cover the paths of least resistance, where it was easiest to walk — for pigs that is. Through the laughter, I could hardly keep the conversation serious; I had to wait for us to calm down.

"If we spot a pig, we try to keep it in sight and whistle loudly to get everyone else to gather round. If we get lost, we stay put and whistle three times in succession until the others find us."

"Unless the pigs find us first!"

"Yeah, maybe when they see their Daddy Michel they'll jump with joy into his arms!"

We left on our mission in different directions, chuckling our way into the woods. After a minute I slowed down, figuring I'd have a better chance of locating the pigs if I could find hoof marks and establish their direction. Scrutinizing the low holes, I finally found their tracks and placed myself in stalking mode, alternately advancing and stopping to watch and listen. Being domestic animals, it stands to reason they couldn't have gone too far. I stopped at the edge of a large clearing where tall ferns abounded, forming a thick carpet of pale green colour. My eye caught a slight movement in the middle of the clearing, where fronds were waving unnaturally. They're here.

When the buddies arrived, I ask them to surround the clearing. We slowly crept toward the pigs from all sides, and watched the tops of the ferns to detect their movements underneath. To no avail. As soon as we approached, they effortlessly zipped between us. Then they regrouped elsewhere under the fern patch, grunting. We tried again, with the same results. Soon we are all chasing down the pigs, attempting to grab them, but we couldn't see them. Five of us cornered one. He passed right between my legs.

Ferns Crack Me Up

I got down on all fours to see under the ferns and gain the pig's perspective. I heard the pigs grunting, without seeing them, and understood they were communicating to find each other. I tried to imitate them and spotted one coming toward me, only to turn at the last minute.

"Hey guys, get down and oink. It'll confuse them and they won't know where to go!"

It seemed to work, as Michel grabbed one by the tip of the leg. But it shook and wiggled so much he had to let go. We were encouraged now and we all went at it harder, each of us grunting and running around on all fours under the ferns. I got up to witness the scene. The ferns were waving all over with loud grunts emitted from all parts. I couldn't tell the pigs from the humans. I broke up, cramped in half with laughter at the ridiculous scene. Then the next guy got up, saw what I saw, and joined me in roll-on-the-ground hilarity. Instantly we all grouped together in glee, holding each other up so we wouldn't fall down, blind from tears.

Back in the cabin, the beer was cold, the pool table fuzzy, and the camaraderie priceless. An hour after we had given up and left the pigs in the fern patch, one of us went outside to notice the jail breakers had returned on their own, apparently satisfied with their escapade and ready to feed. A nice introduction to animal behaviour.

Over the years, nature has become a tender old friend with whom I like to reminisce. Be it wave-splashed boulders, a deep green bog, a trunk overgrown with oyster mushrooms or a bright, furry caterpillar feeding on a leaf, each and every wilderness scene reminds me of an adventure. Whenever I cross a patch of pale green Ostrich fronds, a vividly comical film of a natural human/pig pen instantly fills my screen, which leaves my unwary hiking partners perplexed as to why ferns crack me up.

Seventeen

Exquisite Rain

> "If you are in a shipwreck and all the boats are gone, a piano top buoyant enough to keep you afloat that comes along makes a fortuitous life preserver. But this is not to say that the best way to design a life preserver is in the form of a piano top. I think that we are clinging to a great many piano tops in accepting yesterday's fortuitous contrivings as constituting the only means for solving a given problem."
> — Richard Buckminster Fuller

Spending time in the natural world incites one to become an environmental steward. You can't help it. Something stronger than ourselves compels us to protect those precious places which have provided so much wonder, so much inspiration, so much delight. We simply wish to let other children of the world play with the same soul-filling awe.

My own initiation rites in ecological militantism occurred in my sophomore year as a university student, when I joined Scarborough College's Pollution Probe. The first meeting left me severely disappointed, however, since it was held in a cloud of cigarette smoke. To my purist eyes and nostrils, a sacrilege was being committed. But I was nevertheless hooked by the no-nonsense concrete actions suggested.

WILDERNESS SECRETS REVEALED

An extra shove in that direction was provided by Edouard Blanchette, a high school teaching colleague. Ed had built himself a passive solar house that he could heat with less than half a cord of wood annually. I was impressed, not only by the structure but also by his discourse. Doing more with less. Synergy. Ephemeralization. He was introducing me to the ways of R. Buckminster Fuller, the fellow who had built the incredible Biosphere geodesic dome in Montreal for Expo '67. For an appetizer, I read Bucky's *Operating Manual for Spaceship Earth*. Then came the entree *Utopia or Oblivion*. As a main course I gulped *Critical Path*, and *Synergetics* was served as an elaborate dessert.

Bucky's piano-top-as-a-life-preserver analogy was his way of introducing his general idea that we should always be looking at things as if we were seeing them for the first time. Open your mind. Think freely. Another proponent of this principle was Victor Papanek, whose treatise *Design for the Real World* greatly influenced me to reason differently, to stop blindly following accepted practices. I paid a social price for this, of course, as I arrived at parties lugging a crate of milk and cereal instead of a case of beer and pretzels. From a wilderness survival perspective, though, the MacGyver-style mindset for incessantly observing things in a new light becomes invaluable, and this encouraged my odd behaviour.

For me, thinking out of the box reaped sufficient rewards to make up for the misfit label. As I became involved in a dozen organizations such as Greenpeace, Amnesty International, or Friends of the Rainbow Warrior, I became aware that other so-called misfits shared my concern for planet Earth's future. In fact, many individuals were so extreme as to actually scare me into backing off a bit. I decided it would be better for me to restrict my actions to my own sphere of influence, and I started getting more involved locally. In particular, I founded and presided over STOP (Society to Obtain Peace) and co-founded two research entities, one to introduce sustainable development to the region and another to promote alternative ecological practices. But this didn't satisfy me fully.

Like most young adults I suppose, I was searching my inner self for the contribution I could make to society. I knew I wanted to help preserve nature for future generations, but how? The answer came when I met Sir Edmund Hillary at the International Congress in Outdoor

Education in downtown Toronto's Sheraton Centre in 1983. Hillary spoke to the crowd for an hour and a half. During the first thirty minutes he entertained us with an exhilarating account of his first ascent of Mount Everest in 1953, dressed in leather. Amazing! Then he spent the next sixty minutes in stimulating discussion related to his philanthropic efforts to build schools and hospitals for the Nepalese children. As I spoke to him afterwards, it became evident he was using his fame for a purpose. Motivating.

I arrived in Chicoutimi with a brain-overload headache. Could I too become famous and use my tribune for nature's sake? What would be my Everest? And in the wink of an eye I had decided, just like that. I would finish my doctoral dissertation pronto, and for a graduation present I would spend a whole month surviving in the wilderness. Then maybe people would listen as I recounted my adventures, and I would be able to slot in the message to protect wild places.

That spring, as I dared speak of my project to my friends, I was introduced to Jacques Montminy. Jacques was getting involved with the university's outdoor station in the Laurentide Wildlife Reserve, and his experience as a trapper fascinated me. We hit it off when I exposed my far-fetched idea and he replied that he too had been longing forever for that kind of extended survival experience. We agreed that August should be the favoured month, for then the berries would be ripe and the bugs mostly absent. And he happened to be free that summer. The prospect of spending a month in the bush sure seemed more enticing with a buddy to share my folly. Truth be known, I didn't look forward whatsoever to a solitary journey of that magnitude.

Jacques and I spent a couple of afternoons practising fire by friction by handling the bow together, one on each end. The extra power of this coordinated effort facilitated generating the precious spark. To complete the training for our forthcoming trip, we decided to spend a few days together surviving in the forest. It was the end of May.

"How about next weekend, Jacques? Ready to go for it? We could go from Saturday to Tuesday. I don't teach until Wednesday."

"Sure, why not? But they're predicting a lot of rain!"

"It'll be a good test for us, don't you think?"

"Yep! Meet you here Saturday at 8 a.m. We can stop for breakfast in Laterrière village and then take off into the reserve."

"Sounds good to me."

The wipers were flapping full tilt to clear the downpour from the windshield as I drove toward the university and then to the restaurant on that fatidic day. After steak and eggs, the storm had not abated in the least as we drove onto a muddy logging road toward an arbitrary destination.

"Let's stop when the mileage counter reads ten."

"Okay, that will make it a random spot."

We parked the car and headed out to face the cold and dreary environment. Within seconds, our pants were soaked through and minutes later so were our leather hiking boots. At least our raincoats maintained our core body heat as we sought out a suitable flat spot to set up shelter. But as we walked and brushed by young fir trees, water somehow found a way to seep through seams or to sneak in around our necks. Plus, as we worked our way uphill we overheated and sweat, slowly but surely soaking our base layer. Time for a fire.

"If we get this kind of weather during our month long venture in August, I guess we'll just have to wait it out until the sun shows up!"

"At least it won't be this freaking cold in August."

"Hope so!"

"No use suffering needlessly now. Let's use my magnesium fire starter."

The nasty deluge tried its best to bully us, as if accusing us of taking the easy way out. I peeled a couple of sheets of soaking wet birchbark from a tree and split them into thin sheets to access the dry layers inside. But the rain dripping down my sleeves inundated them instantly. Jacques came lugging a substantially wide piece of bark he had ripped from a downed and rotten tree. He held it above me as a temporary umbrella so I could achieve my goal. With my pocket knife, I scraped miniature shavings of magnesium into a thumb-sized pile. I then tore a thin sheet of dry inner bark into spaghetti-like strips, which I delicately positioned above and beside the magnesium tinder. The metal striker was a rod of ferrocerium that was embedded into the piece of magnesium. To avoid

Exquisite Rain

scattering the pile of shavings, I ignored the standard instructions that called for hitting the striker with the knife blade. Instead, I held the perpendicular blade steady against the striker just above the shavings and pulled back hard on the ferrocerium. A bright spark shot into the magnesium tinder, which flashed into a flare, lighting the bark. Tons of bark was added to the capricious flame in order to generate sufficient heat to

Example of a birchbark umbrella.

WILDERNESS SECRETS REVEALED

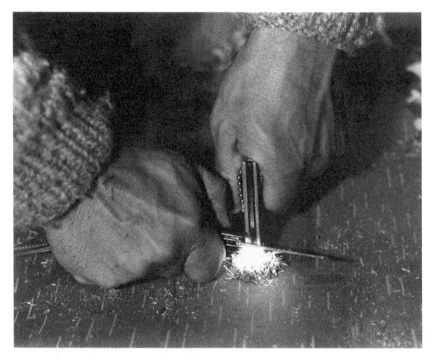

Above: *ferrocerium/ magnesium metal striker.*

Left: *During rain, the fire must be kept huge until a bed of coal forms.*

dry out dead fir branches. The secret to keeping a fire going in the rain is to pile on the wood, this way the top logs act as umbrellas for the lower ones. The fire must be kept huge until the top pieces dry out and a bed of coal forms underneath. Only then can the vigil relax. By that time, the logs are usually cut in half, simplifying further manipulation.

Jacques and I fetched more firewood, avoiding birch except when the bark had been partially ripped off as the tree fell. The bark of a birch, and to a lesser extent that of cherry and alder too, acts as a plastic bag, preventing the wood from drying. This means that birch wood normally passes from green to rotten without ever experiencing a dry stage. Whereas split and dried birch makes great firewood for stoves and fireplaces, this essence is next to useless as fuel in the wilderness. Soggy birch logs, however, provide great building material, and we piled up a few as a windbreak. Birchbark still intact on these rotten logs also make great shingles for covering survival shelters, and we carried heavy stacks of them back to our fire site.

A hard worker, Jacques is. By lunchtime, we had gathered impressive cords of wood and a semblance of a shelter structure had been built. We were still wet, but warm from exercise. As nightfall approached we were satisfied with our lean-to and sat on a large log underneath drying out our clothing and boots. The relentless rain and pitch-black surroundings confined us to our humble abode.

We whiled away the hours by chatting and exchanging past experiences, getting to know each other. Sleep seemed unlikely on the soaking-wet bough bed. I attempted to dry some of the glistening evergreen branches by holding them next to the fire. The rain saturated them as quickly as they dried, and if I tried to speed up the process they caught fire, throwing countless sparks into the air. So we sat sideways on our log, leaning on each other's backs and waiting for the new day. In the wee hours, a substantial downpour revealed a great many leaks in our shelter, and we were glad when it eased off. Yet, it sure was awe-inspiring to witness the raw power of nature's dramatic spectacle from our front-row seats. We dried our pants again.

The poor weather continued. The forest was drenched, as if a giant hand had turned it upside down and dunked it in a lake. We drank easily out of a birchbark cup held at the edge of our roof. We were starved, but nature's grocery store was closed and locked tight. To our surprise, our three-night wood pile had already been largely consumed. We had to get wet again. Working with a tenth of the previous day's enthusiasm, wood had to be logged from an ever-increasing distance. Most of our day was used up sleepily doing that and adding another layer of bark shingles to the lean-to. By the time we had scorched our clothes and boots to dry them one more time, we happily noticed that our bough bed had finally turned from wet to damp. Yet, cold and humidity by and large prevented sleep on the second night too, save a two-hour lull in the rain that provided some respite from continuous tending of the fire. With such low pressure, smoke also became a major irritant.

When dawn appeared, we were greeted by the pretty, wispy fog and misty rain that emphasized the rich deep colours of the pitter-patter world around us. Even through bloodshot eyes, we could not fail to admire this reward for being there. But grim reality set in as we noted no change in weather. All day there was rain, rain and more exquisite rain. Enough to magically propel me into the world of fish. To break the spell I poked the fire and replace the burnt ends into a teepee shape to coax the smoky embers back into a cheerful blaze. We huddled close to dry our gear for the hundredth time and hang out for hours, half asleep, too weary to budge much. Like two prisoners in a dungeon, we simply existed, praying to be delivered from our torture by a sunny friend. No such luck.

We were out of wood, and the quest for logs resumed. With heart-wrenching effort, we gathered up a dozen each. Totally worn out, drenched to the bone, and shivering, we were well aware that that night would be the most horrible yet. The lack of food weighed heavily on us. It was almost dark.

"Hey Jacques, how about a big fat juicy steak?"

"With french fries?"

"Yeah, and chocolate cake too."

"Let's get the hell out of here!"

Exquisite Rain

Years later, I discovered this poetic gem by Langston Hughes: "Let the rain kiss you. Let the rain beat upon your head with silver liquid drops. Let the rain sing you a lullaby." Those are brilliant words, Mr. Hughes. I can relate. For in spite of the hardships rain brings, it's rich inherent beauty remains as my only souvenir.

The unsuspecting waitress was amazed at the appetite of the two charcoaled-faced grubby guys when she brought out the order. As I was by my partner, who had shone through the waterlogged ordeal without a twitch.

I was elated to have discovered in Jacques a trustworthy partner for this outlandish month-long stint in the woods. We shook hands for a first of August departure.

Ready…

Freshly energized by this commitment from Jacques, I disappeared into a logistics black hole, so busy parsing details that before I knew it we were dumped on the doorstep of that much anticipated day. But I felt we were ready. Our Survivathon agent, Jean-Claude, had organized the airport media show around the helicopter that would whisk us away. Our scientific gear was well-packed in the canoe drybag, our radio batteries were fully charged, and our bellies were stuffed with food.

Set…

Okay, perhaps half of that stomach content was knotted tension. We were to face the northern boreal wilderness with just the few clothes we were wearing: socks, hiking boots, jeans, belt, t-shirt, dress shirt, light wool sweater, unlined jacket, bandana, and hat. Our pockets contained no more than a next-to-useless wallet with money and credit cards, loose change, and a key ring. Nothing else, not even a matchstick. Scary. I cradled my two cookies as a child does his security blanket.

Waves of hands cheered us on as we approached the helicopter with knotted throats. We were four on board: the pilot, a videographer, Jacques, and myself. The motor roared.

Go!

Eighteen

The Charcoaled Commandos

"When you get into a tight place and everything goes against you, till it seems as though you could not hang on a minute longer, never give up then, for that is just the place and time that the tide will turn."
— Harriet Beecher Stowe

Emotion was the fifth passenger in the helicopter as we flew over the dark waters of the Saguenay River and the majestic Valin Mountains on our way to the 50th parallel. I was overwhelmed with doubt. The unknown was that I'd never visited the boreal forest so far north. Will the resources be familiar? Will I succeed in lighting the fire so essential to our survival in this rainy weather, particularly without tools? The fear of failure gnawed at my innards. What will people think of us should we quit tomorrow? In the space of an instant, I regretted not having hidden matches. Too bad, it was too late now.

Fog forced us to land in the village of St-Honoré. The wait was interminable to me but Jacques took it in stride, calmly chatting with the pilot. Didn't he realize that every wasted minute left less time for rubbing sticks and setting up camp? It was 10:55 before we took off again, which meant we wouldn't get there before 12:30. I was already feeling hunger pangs!

The infinite swatches of clear-cut logging below added to the churn in my stomach. But soon an invisible paintbrush swept a deep green colour onto the scene. To the edge of the canvas in all directions lay spruce trees, interspersed with baby-blue lakes. The fluffy cumulus clouds just above us fascinated, providing the illusion we were in a spaceship being propelled through a thin corridor of void.

We were pleased to note that most of the small waterways were inhabited by one or more beaver huts. Would the beavers fall prey to our rudimentary traps? Jacques assured me they would. I checked our waterproof bag that contained our scientific equipment. Yes, our special trapping permit from the Ministry of Natural Resources was in there.

More than forty-five minutes earlier the last trace of civilization had disappeared. A huge and mysterious forest awaited. Grandiose!

Finally, shimmering on the horizon appeared the tiny heart-shaped lake where we would begin our journey.

"I think I can land there, on the small mound of bog sticking out."

"Can't you bring her down on the other side, across the creek? Looks swampy on this side, and there are no trees."

"Sorry, but there isn't enough room over there."

As our experienced bush pilot descended, I was already scanning the surrounding area to identify edible plants and species of trees available for fire. No cattails nor cedar, damn!

"I can't stop the engines, the ground is too soft!"

"That's okay, we'll jump out anyway."

As had been agreed upon prior to departure, our cameraman Gilles exited first. As soon as he cleared the propellers, we took off and circled the lake again. We saw Gilles below shouldering his camera and filming our second descent. In the absence of cedar, I was desperately trying to spot a dead aspen or fir, my second and third choices for fire-starting material. Too late, and I dared not ask for another chance. In the distance to the north, I did catch a glimpse of a greater lake.

As we landed for the second time we saw Gilles knee-deep in water. In a flash, we removed our boots, knowing the difficulty of drying them by fire. So we hopped out in bare feet, and the Survivathon began!

The Charcoaled Commandos

We stooped away from the helicopter blades. If an adrenaline meter existed, it would surely have popped. I felt so small, so far from everything, so helpless. We petitioned for adventure, and here it was!

We met up with Gilles, loaded with our heavy radio, its unwieldy antenna, and the rest of our scientific gear. He screamed above the deafening roar of the engines: "You'll die of blackfly bites up here, you guys! I'm already bitten beyond belief. Quick, let me grab the bottle of fly dope in my jacket on the seat of the copter!"

Indeed, we were already being attacked by an extraordinary number of blackflies that seem delighted to taste fresh city meat. We could not resist the urge to beg a little repellent from Gilles. We justified our request:

"After all, Gilles, a guy lost in the woods would at least have covered himself with dope before leaving!"

We slapped on the insect repellent. Thickly.

"Hey, half the bottle is gone!"

"Sorry, Gilles, we need it more than you do!"

Jacques and I knew that the meagre protection would only last a few hours. If only we could light a smoky fire before the repellent wore off! Never in a million years had I imagined encountering so many blackflies in August; in the south, they're gone by mid-June.

Gilles, who now sought refuge in the helicopter, threw us a word of encouragement before leaving:

"I wouldn't stay here with you guys for all the money in the world!"

Our last contact with civilization zoomed away from the flies, away from us too. They were but a dot now, another black spot in the swarm. Gilles last sentence still resonated. Did we bite off more than we could chew? Panic squeezed me tightly, but I strove to regain control; after all, I always felt this way at the beginning of a survival experience. In such circumstances I reminded myself that I could surely continue for at least an hour. Usually the situation improved.

An overwhelming silence descended upon us. We really were in the middle of nowhere, the only men on a new and empty planet. Mother ship, please beam us back up! But it was too late, we'd swum into our own month-long net. We'd fallen in the four-week pit we ourselves dug. We'd stepped in our own thirty-one day trap. Enough! I must concentrate on my one and only immediate concern. Fire! Damn! I absolutely must succeed in lighting a fire! It wouldn't be easy without matches nor tools in this forest dampened by recent rain.

It seemed the news of our arrival had quickly spread throughout the blackfly colony. Their number had quadrupled since the helicopter blade had stopped flattening the nearby vegetation. Hundreds and hundreds of the blood-sucking pests teemed on my pants, and I could almost cut with a knife the swarm that revolved around my face. Never, in all my stays in the forest, had I ever encountered anything so demoralizing. But the diethyltoluamide repellent still prevented them from landing on our vulnerable urban skin; unfortunately, it wouldn't last long.

I stared at Jacques, barefoot in the water, boots around his neck, radio in hand, with a thick cloud of insects floating around him. Emptiness. Not knowing where to start, I questioned him with a look of despair. But his eyes were interrogating me in the exact same way.

The lake seemed so much larger now that we stood on its sloppy shore on tender feet. Our ankles were already decorated with dozens of swelling bites, so we headed out of the bog and donned our boots. Aiming left toward the creek, we intended to go uphill toward a few balsam poplar trees poking out above the black spruce.

A few steps further another disappointment awaited us — the blueberries were green and tiny, nowhere near ripe. The Saguenay Region is the blueberry capital of the world, and I had imagined stuffing my face with juicy clusters. To make matters worse, we saw no signs of animals, not even a hare pellet. Fear of hunger shrivelled my courage.

The muddy creek turned out to be wider than we had imagined from the air. Following it upstream, we eventually found a huge log that could be toppled end-over-end until it fell to create a wobbly bridge. Huffing and puffing, we paused for rest and I entrusted my first notes to my precious voice recorder:

"Survivathon, August 1st, 12:45. We've landed in a typical boreal forest composed mainly of black spruce, with some balsam fir, poplar, and birch. Sphagnum moss covers the ground, prime habitat for blackflies. There's an army of those monsters, we can hardly see our pants. I've noted very few edible plants, only Labrador Tea and a couple of unripe berries in the *Amelanchier* genus. I did hear the purr of a red squirrel though."

My companion in misfortune imitated me, recording as well:

"August 1st. The flight in lasted about an hour and a half. We were happy to see them go. We've walked around the lake for thirty minutes or so. There are tons of mosquitoes and mostly blackflies, it's scary, horrible. I hide my hands in my pockets and they get through the seams. This is hell, especially when we're walking in the bog. The beaver huts seem uninhabited. I've seen tons of moose tracks and some partridge droppings."

As I gathered the materials for fire-starting, Jacques was tampering with the radio, attempting to set up the antenna between trees. He couldn't get through. Without news, Jean-Claude had spoken of five days before rescue. That's a long time in those bugs, especially without fire. And we may have been in deeper trouble still if cold torrential rain was added to the equation.

My fire-starting endeavour wasn't going well. Not well at all. In fact, after a couple of hours I was back at square one. The bow was too crude and heavy to build up friction speed, the crooked spindle kept popping off the slightly rotten board, and the drilling had worn a hole straight through the handle.

A sharpened key resembles a scraper more than a knife. With this shabby two-centimetre blade, it took forever to straighten out another crooked branch to make a decent rubbing stick, for I couldn't find a straight piece of wood anywhere. It was a work of patience and frustration, especially with the hungry flies tirelessly pursuing us. Then the spindle cracked while sharpening the ends on a rock. I tried it anyway, but it broke in half as soon as I started drilling.

On the third try, the shoelace tied as a bow string snapped just as the smoke started billowing out. Jacques handed me one of his. Now the too-short spindle needed replacing again. To make matters worse, the repellent was wearing off and we were scratching our first bites. Our discouragement was total.

The moment was intense. I laughed. Nobody would ever believe how many loathsome bugs were attacking us. I pulled out a pen from the scientific kit and drew a two-centimetre square on the thigh of my pant leg.

Now I had a scale to measure the density of the flies. There were six of them crawling in that miniature arena. Unreal!

I prepared still another spindle, lucky to find fir that time. It gave me hope as it spun freely. Cutting a notch in the board with the key was the toughest part of fire-starting. Jacques was sharpening his belt buckle to speed up the process.

Finally ready, we grabbed the bow together and resumed drilling. I was entirely inhabited by doubt, but then welcomed smoke poured out. No; failure again. We had cut the notch too wide and the spindle flew out. I counted three mosquitoes and five blackflies in my two-centimetre square. We performed the see-saw motion one more time but drilled right through the board without obtaining the desired ember. I fetched a thicker plank of wood from a fir stump.

High voltage shot through my panicked nerves. The repugnant sight of Jacques's swelled ears didn't help. Never have I started a fire under such arduous and technically difficult conditions. But we must try again. What's that old saying? "When the going gets tough, the tough get spinning." The wood knot I was using as a handle was also wearing through. I grabbed a ten-cent coin and forced it to the bottom of the hole by nudging it with a pointy stone; this would act as a ball bearing.

Twilight woke up millions of mosquitoes who joined the blackflies to devour our hands and face. Their buzzing was exhausting. At least the fourteen blackflies in the square were silent! We were sweating like pigs inside our anoraks, especially after the intensive effort needed to swing the bow. Nauseating.

With blistered fingers I surrounded the spindle with the bowstring, and like a machine settled back into the drilling position with Jacques. We needed this fire at all costs! I changed my mind set: I was a tough caveman, a primitive, an animal even. I felt no pain. A transformation took place. I was empty of all other concerns, like a shaman before his sacrifice. Blackflies, mosquitoes, sweltering heat, or blisters couldn't stop me. I had become a savage beast struggling for its life. I yelled with all my strength while spinning the drill so quickly it melted as in an out-of-focus photograph:

"PULL, Jacques! Let's do this!"

More and more pressure on the spindle, lightning speed on the bow, sweat on the forehead, smoke in the face! STOP! I gently removed the spindle, trying to control my coughing, for smoke had entered my throat, eyes, and nostrils. Lactic acid cramped my muscles.

"Jacques, look!"

In the notch, a tiny trickle of smoke arose. An ember? Yes, an ember! But how many times had I managed to obtain such an incandescence without being able to ignite it? And this one was so very small, barely larger than a pinhead! Trembling, I picked up the ember and fanned it with my hand, for even the humidity of my breath would be enough to put it out at that point. Gently, ever so gently, I added the handful of powdered birchbark I had stuffed into my pocket. Now I blew softly, so as not to scatter the tinder, yet enough not to lose the frail ember. The smoke seemed to decrease. But no! It was gaining. I blew harder. And harder. And harder still! There was smoke galore now. Almost there! A powerful last breath! Flame!

I dropped the fire burning my hands and rush to add piles and piles of birchbark.

"Quick buddy! Help me find dry branches!"

We added mountains of sticks, and soon the fire roared dramatically. We couldn't contain our joy. We yelled like crazy. I pulled out my voice recorder, laughing and yelling like a madman:

"YAAAAOUOUOUHHHH! Fire! Son of a bitch! It's 17:56 and we finally have fire! We're happy campers now! YAAOUOUHH!"

I took Jacques by the arm and we did three rounds of a square dance. We jumped, sang, and shouted like savages. It would have been funny to watch. If there were animals in those parts, one thing was for sure, they were gone.

Even after thousands of times, I had forgotten how good it felt to watch the forest glow. Comfort, relief, quiet. I was neither hot nor cold. Flies didn't irritate me anymore, my blisters became painless. I was no longer an animal, but rather a human being who had managed a significant part of his mission, one that would allow him to survive. I was reconciled with my friend, Nature, who had invited me to enter her home and light her fireplace. It was now possible to live there a month. I sat down, in adoring admiration of the simple flame.

A dry throat woke me from my peaceful reflexive moment and I was face to face with harsh reality again. Dehydrated, we slipped down the mountain, crossed our log bridge, and tramped down to the lake for a sip of hard-earned water. As we climbed back to our fire site, we took stock of our situation, which wasn't pretty — no food since breakfast and nothing edible in sight, no firewood gathered nor shelter built, a damp and cold environment, and relentless flies torturing us. My forehead was bleeding, Jacques told me. He showed me the red welts from bites under his belt. Those nasty creatures managed to infiltrate every tiny crack in our clothing. Dreadful.

Finally, we could turn our attention away from the fire. To think of the work that could have been avoided by a simple paper match! Never again will I enter the forest without a large supply. Never!

By spreading the long wire antenna on top of bushes near the lake, Jacques finally managed to obtain a 1-out-of-5 communication with our radio host. He informed the audience of our success with fire, so there was no need to worry about us. Tomorrow, we would no doubt be able to install our transmitter in a more appropriate location.

Like two camels, we filled with water and trotted back up to the fire that was yearning for fuel. We revived it and then sought substantial pieces of rotten wood to smoke out the beasties. We were the ones getting smoked out now; we couldn't find the equilibrium.

In the boggy mess, we couldn't find a suitable place to sleep. Night was falling fast, so I hastened to search the territory south of the fire while Jacques scrutinized the north. Nothing striking our fancy, we had to make do with a metre-deep mini-cavern that would cover our legs should it rain. At least the soil was dry in there. With a bit of landscaping, we judged it would suffice as a temporary shelter.

I transferred the fire to the camping site using a fungus in the genus *Fomes*, which grows as shelves on rotten tree trunks. Once placed in the fire for a while to start the combustion process, this gift from nature smoulders continuously while it consumes itself. With the fire crackling in the new location, we hustled to build a log wall to protect us from the rising wind and break some fir boughs for our manger. I was afraid our lumpy mattress would be rather thin. Darkness was falling upon us like a menacing guillotine, and not a piece of nighttime fuel had yet been picked up.

We sought the precious firewood with alacrity to ward off the damp coolness of evening, which had penetrated our opened-windowed wilderness home. As usual, we fetched long logs from a distance, ignoring the close-by "no touch" zone reserved for night time gathering by the glow of the fire. Precaution was taken to prepare a pile of dry bark and twigs to relight the fire should the need arise during the night. Then we each scrounged up a huge piece of birchbark torn from a rotten tree to serve as an umbrella in case of a storm. We finally crashed, empty.

But our fate took a turn for the worse as we sat immobile by the fire. The countless insects seemed attracted to the flame and added to our existing bites, which itched furiously. Mosquitoes, heartless, attacked even our eyes, which hopefully wouldn't swell shut. It seemed impossible to settle for the night because of these horrible creatures of the swamps. I inserted branches in my pants to create space between my skin and their stingy darts. We suffered scratchy bites in our belly buttons, around our hairlines, ears, ankles, and wrists. But at least we had fire. Thank God!

In fetal position, we tried to take advantage of the beneficial heat of the fire, but it cracked and spit violently, having been fed only spruce. The embers flying onto our clothing forced us to remain vigilant. In spite of all this, we found ourselves pleased with our first day. Panic had disappeared, we owned fire, and we were lying down, albeit crookedly.

My precious tape recorder rested near my log pillow. It was greatly motivating to think that my data might serve science. I tried to sleep regardless of the buzzing music taunting me. A wink later I vented my frustration to the electronic psychologist:

"20:55. I'm already awake and standing with the fire almost out. Mosquitoes surely don't know what night is for. They sting even through my pants. Our beds are miserable, and the thermometer has dropped to under 10 degrees Celsius, much colder than I had expected. Maybe it'll get rid of the flies? We're under the impression that every insect in the forest has free access to our skin. Will we resist panic? Tomorrow, we must find a windier spot with less peat bog. We've got to get out of this infested area!

"22:26. We've almost lost our fire again. I woke with a start to find that there remained no more than a few tiny embers under the ashes. We shiver from the cold, and even more at the thought of losing the fire. All the problems seem to have been mailed to our address. Morale is flying as low as the flies.

"12:18. I stoke the fire.

"1:00. Once again stoking the fire. Long night!

"2:06. I have to get up again. The cold is now intense, but at least the flies have disappeared. Jacques says he's been playing with the fire all night too.

"2:15. Impossible to sleep. We're seriously freezing; it must be near zero degrees. We'll check the minimum-maximum thermometer in the morning. We obviously didn't gather enough wood.

"2:35. Frozen feet, I decide to sit a few minutes to dry my socks.

"3:03. We're out of wood, must go fetch some in the pitch-black darkness. My teeth are chattering.

"4:55. Can't sleep more than twenty minutes at a time since I must continuously tend the fire. Feeble light on the horizon; another tough day coming up! I'm going to toss all the remaining firewood on the embers and try to get at least a bit of shut-eye.

"Just woke up scratching myself furiously. What a nightmare! I dreamed I was eight years old and had discovered an ant nest under a rock. They were climbing all over my body, biting me to death as I ran, panicked, all over the place. Reality isn't much better."

I glanced at my watch: 7:37. Gently, I rearranged the fire without waking Jacques, who was lying there looking oh-so-miserable. I can't believe I just suffered the same fate too, on a thin carpet of bumpy branches, with my feet higher than my head. The minimum-maximum indicator showed that the temperature dropped to 1°C. A single degree from freezing! No wonder we were so icy cold. And no wonder my boots were still so stiff. Jacques woke up.

"Hey, good morning! Did you sleep well?"

"Fantastic! Better than at home, thanks. And yourself?"

"To be honest, I froze a bit. Spruce flares too fast, and rotten birch smokes without burning."

"I know, and it was hard to sleep with this handkerchief on my face. But that's nothing. I feel like I have crusty bites on top of my crusty bites."

"Yeah, me too."

"I'll try to phone the station and let them know we're still alive."

"Okay."

Already two blackflies were crawling in the square drawn on my pant leg, part of the cloud of unwanted friends who had come for breakfast. In desperation, I spread fir sap all over my face, neck, and hands, in hopes that it would act as a varnish to protect me from the bloodsuckers that were driving me insane. Worse than glue, the sap ran onto my clothing, my beard, and my hair. So I rubbed black dirt onto the gooey mess to sponge it up. I bet I looked awesome. No matter, I'd try anything that might help in the least.

Time to leave the oversized insect incubator. We had to find a windier and less humid camping spot. We decided to follow the muddy creek to the pond we sensed beyond; it would surely lead to a lake.

Carrying our fire became the new challenge, since we could no longer find any shelf fungus in the vicinity. And without cedar bark, I couldn't prepare a fire bundle either. We tried to make do with a rotten stump that had been smoking for a while. Before leaving, we scraped a large area around the fire and piled on some dirty logs to keep the coals alive for an hour or so.

We soon arrived at a pond of stagnant water, like black ink. No beavers. We continued. Our boots were soaked, but we couldn't care less. Labrador tea and other tangled scrub slowed our progress. Further on, to our great satisfaction, we discovered a cigar-shaped lake and gained its shore. But watch out! The embers on our stump were dying.

"Quick Jacques, bring me birchbark!"

"Won't be long!"

Desperately, we tried to tease a flame from the last shy ember. We blew, blew, and blew, but it was too late. Damn! The tiny reddish glow paled and faded away. Nothing. Nada. Zero. Zilch. Unless there were still embers at the former camp! It had been well over an hour since we'd left. No use both of us regressing, so I headed alone back the way we came. I couldn't imagine mustering up enough strength to repeat the rub-the-sticks trick; that set me running. On the way good fortune delivered a *Fomes* mushroom.

Phew! I'd arrived. Fire? Yes! A trickle of smoke rose gently from the grey ash. Thank God! I coaxed the buried embers into flame with bark and twigs and added the fungus on top. While it dried out, I broke a Y-shaped twig of birch to make a clamp of sorts to carry the now glowing fungus. I returned to meet Jacques, not too proud of this dramatic waste of time and energy. At least he had found a couple more fungi.

Reassured by the security of our resuscitated friend, we continued our hike around the lake toward an imagined campsite. We happened upon a pool of stinky water out of which some green and slimy frog eyes were peeking. Tricky creatures; they hid in the mud and outwitted our half-hearted attempts at snatching them. We resumed our journey.

While Jacques lit new fungi and squeezed them between forked branches, I climbed halfway up the nearby hill to gain a better point of view. The lake was entirely surrounded by stunted brush, allowing a windy space to the east, but without firewood.

The scorching sun bullied me into dragging my feet. Heat exposure was a concern, but I couldn't take off my anorak without inviting to lunch my personal cloud of flying leeches. In fact, I had resorted to pulling my t-shirt up over my head and tying a knot in it to reduce their entryway to a tiny opening, just enough to see and breathe. Yet they still managed to sneak by and bite my face.

The top of the hill levelled off to a kind of wavy plateau. In the distance, to the north, I spotted the reflections from a huge body of water. I ventured deeper into the forest. The vegetation there was more encouraging, with less moss and abundant firewood. Another plus: I felt a slight breeze. On the other hand, we would have to climb down the mountain whenever we would need a drink.

The slight movement of a Ruffed Grouse set me into hunting mode. I kept it within my field of vision as I tied my shoelace to the end of a tall skinny pole. The chase became drawn out and painful as the flies profited from my forced immobility while I maintained my stance during the approach. Once done I returned to Jacques, who was sitting by a smudge fire. I showed off the results of my hunt, or rather lack thereof.

The Charcoaled Commandos

Dressed against biting insects.

We finally opted against camping up the hill and headed north instead. In our hands, our warm and smoking friends; on our backs, the radio and other scientific paraphernalia; in my head, the frustration of having muffed-up securing our meal. Jacques encouraged me. There would be other opportunities.

What a contrast to last night's freezing temperature! The sun was beating us into a mushy pulp. We moped along, sweating profusely for two interminable hours until our mushroom tinder ran out. While I started a fire, Jacques headed out to find more fungi, but returned empty handed. We attempted to move forward with bundles of sticks, but had to ceaselessly stomp out the falling mini-meteorites that threatened to set off a forest fire. We could only proceed a hundred metres at a time. Frustrated by the slow progress, I barged deeper into the forest to rummage for some more polyporus fungi, and finally spotted some, in the

Ganoderma genus this time. Filled to the brim with hope, we finally drew near the mysterious lake.

Jacques veered off to the right while I headed to the bay on the left. But the longed-for lake disappointed. With a shaky voice, I vented my emotions to my recorder:

"12:29. The shore of this large lake is worse than the place where we slept last night. Its marshy shores are a paradise for mosquitoes and blackflies. I'm completely discouraged. Flies are as fierce as always. Everywhere I scratch myself I find patches of crusty blood. Clouds announce thunderstorms, we have no shelter, no firewood, and nothing to eat. We are dead tired and our boots are soaked right through. Our morale hangs on the slight thread of our portable fire. I feel at a loss in this inhospitable environment. Will we be able to cope for thirty-one days in these conditions?"

I shook myself after the futile exploration. I tried to catch up to Jacques but he was nowhere to be seen. I shouted. No response. I was sure he had headed this way. Eastward, I clambered up a hill to bypass the swamp. Another yell. Nothing. Shit! More lost energy. I pursued my exploration toward the great expanse of water on the other side of the knoll. Nothing. I bellowed out his name as loudly as I could. Finally, a weak answer. Why did he go so far? I finally caught up.

Was I seeing things? A mirage? The shoreline there was lined with tons and tons of driftwood. A gift from heaven! Oh, what a comforting sight. Jacques thought we must have been on one of the northern arms of the huge Pipmuacan Reservoir. Before I could seek out a suitable camping spot, my pal insisted on setting up the radio antenna. He seemed overly preoccupied with communicating. I nevertheless humoured him and scurried up a branched spruce tree to attach one end of the wire. A huge tripod was erected to hold the other end. Nearly an hour later, the operation was complete, but to our dismay we heard nothing but static.

At 3:00 p.m., weak and weary, we finally decided to set up our permanent fire against a giant overturned stump that lay about fifteen metres from the waters' edge. A long pole leaned against the stump above the fire would serve as the framework for a triangular lean-to. Sleep deprivation caused us to work haphazardly and with muffled enthusiasm.

The Charcoaled Commandos

Nevertheless, after a couple of hours we were encouraged by a semblance of shelter and huge piles of fir boughs waiting to become a soft mattress. Several large sheets of birchbark were ready to be laid like shingles on the roof of our future home. Even the insect monstrosities seemed to have less appetite there, as I only counted two in my small square. I took a breather on shore, where the welcomed wind chased the flies away somewhat. Pretty view. I sipped beautiful clear water, dumbfounded that so-called civilization pollutes waterways to the point where this is no longer possible. Weird.

The driftwood takes on all sorts of grotesque animal forms. Waves sweep across the sandy shore with a melancholic slip-slap, lending ventriloquist voices to the beasts I invented. My attention turned with concern to a sky darkened by monstrous cumulus clouds. My reverie was abruptly interrupted by Jacques's worried scream.

"André! Come quick! The stump is on fire!"

I ran to the shelter. Three-metre flames soared through the air. I rushed to help Jacques throw sand on the blaze. To no avail! Stunned, we looked at each other, speechless. Soon, the stump's intersecting roots burned up, and the earth wall on which our ridge pole was supported collapsed, along with our entire shelter.

"I don't get it, Jacques. I've built dozens of fires against overturned stumps like this in the past and this has never happened before."

"Well there's nothing we can do about it. We'll just have to start again from scratch!"

The weather cooled. The wind moaned and deep grumbles could be heard in the distance, even over those in our stomachs. Would it rain? Our protection was gone. Smoke rose heavily from the stump, even once the fire was out. It was black and ugly, real ugly. I was dirty, frustrated, and disgusted. What folly it was to come here to eat misery. To think I could be home, lying warm in my cozy bed.

Morale at zero, I strolled the shore with cement feet in search of a new shelter location. But the ominous horizon scared me back to the fire, our only refuge. We decided to build a second fire two metres away from the first and installed our bough beds between the two. We each grabbed a birchbark umbrella and anxiously awaited the towering storm, announced by the gloomy darkness and renewed vigour of the countless miniature vampires. In spite of all of our energy expenditure that day, we had nothing to show for our efforts. We worked unfocused, pushed by the mood of the moment, our minds confused and ineffective. What would the future bring?

I woke the next morning with the impression of having just napped a few minutes. No matter how I writhed and wiggled, I couldn't attain a comfortable position. At least we survived the intense but short-lived downpour that night, even though we had to waste a couple of hours drying ourselves afterwards. Jacques was still snoring away. His ears looked like cauliflowers. I got up and stoked the fire for him.

My mouth felt thick and pasty. My sooty clothing disgusted me. Plus, I stunk like carrion. I could at least brush my hair a bit! But the comb refused to scrape through my horse's mane. I headed to the lake to splash water on my face. Then it was time for the daily spruce-gum chew, accompanied by grinning from the bitterness.

It was time to work; the flies were still asleep! I swore to myself that I'd be sheltered by nightfall. I scrutinized the environment for a new lean-to location, without being able to visualize any. Within minutes, dew had soaked my boots again. Depressed, I returned to plop down on my flattened boughs. Nearby, I noticed the old stump still exuded a wisp of smoke. I remembered reading somewhere that latent embers can spread along roots and cause forest fires. We would have to monitor that.

As I casually wandered a few steps south from my tragic bed, I noticed a tuft of tightly growing fir trees, four of which formed a circle. It gave me an idea. My right boot lost its shoelace for a second time. I cut a short piece by smashing it between two rocks and used it to retie the top two eyelets of my boot. The other end of the lace would play a much more important role. I called Jacques, who had been attempting to establish radio communication non-stop since he arose.

"Hey Jacques! Come help me please!"

"Sure. Just a sec!"

"Look, Jacques, if we tie the tips of these four saplings together, it would form a conic structure upon which we could build a teepee."

"Might as well give it a try. We have nothing to lose."

Jacques pulled the trees together and I tied them off. Then I went inside the three-metre-wide circle and broke off the inner branches. On the solid frame we leaned a few poles and soon the teepee took shape, boosting our morale. Onto this humble foundation we piled long grasses and sheets of birchbark, concentrating our efforts on one side so we could at least enjoy a waterproof sitting space. On the shore I found a spade-shaped root that served to dig a fire trench in the middle of the teepee and out the door. The pit would hold the long firewood and help draft the smoke upwards and out through the metre-wide opening at the top. On either side of the fire we would construct our beds, with enough remaining space behind us for tinder and scientific gear. As I excavated down to mineral soil, tossing out the sphagnum moss layer, I discovered where the blackflies hung out; my anorak resembled a peppered egg. I transferred a few coals into the hole to smoke them out and continue my task. Once done, I lined the pit with rocks. We enlarged the fire and tossed on a couple of long logs that we had to step over to enter the hut.

Hey, it's not so bad in here. The darkness substantially reduced the onslaught of the Lilliputian enemies, a magnificent fire warmed us, and our sheltered corner offered security. Now we had to regain our strength. After three days it was time to find something to bite into. I'd sworn not to touch the two delicate cookies in my pocket for at least a week.

Fresh fish sure would have been nice. I lay in the shelter and began the boring task of rubbing a key on a rock to produce a point. I fantasized, recalling fisherman stories where huge northern pike attacked anything shiny. Rub rub rub. Getting there. Rub rub rub. Then I heated the key and managed to bend it by wedging it in a log's crack. A crude hook was born. I tied it to a shoelace lengthened by my anorak's drawstring. While heading to the fishing spot, I broke off a slender pole and attached the line to it. I nonchalantly tossed the key in the water and pulled it toward me over and over again, like an automaton, daydreaming. What the hell was I doing? I'd never catch a fish with such a short line and without bait! And even if a fish did bite, there was no way this coarse point would hook it! It seemed like my bird brain couldn't reason any better than that of the fish. I must have been more tired than I thought.

"Hey, Jacques, I tried fishing. The fish laughed at me. I heard some frogs jumping into the water noisily, but I couldn't see them."

"Most fishermen get skunked even with modern gear, don't forget. By the way, your lips are swollen as if you had been hit by a boxer, and they're bleeding too."

"Damn flies!"

After a short rest, I burrowed into the forest with the firm intention of finding something to eat. My efforts were rewarded by the discovery of three handfuls of nearly ripe *Amelanchier* berries, those memorable little blue pears of my tender years. Along with a few skunk-smelling and bristly wild currants, the scale tipped at ninety grams, forty-five each. Better than nothing, I supposed.

Jacques returned from a long walk with an equally long face. He couldn't believe how few animals frequented those woods. Other than bear and moose, that was. He seemed to be a competent trapper. I was really looking forward to learning from him.

My camera clicked memories of my vacation there. However, I didn't expect wonderful results with such a dirty lens. Same went for my eyeglasses; they were terribly grubby and greasy and I had nothing to wipe them with. But I could see enough to add grass and bark to the shelter. Then hunger enticed me to go berry-picking again. I came back empty-handed. I stirred the fire and entered the world of dreams for a while.

The teasing heat and joking flies amused themselves by waking the bearded guy. I got up and went off to sneak into their territory again, but faced millions of them waiting there in ambush. With a digging stick, I unearthed some Wild Sarsaparilla roots, a supposedly marginally edible plant that I'd never experimented with. I scraped until my pockets were full. Surely, the plants contained carbohydrates; they would no doubt provide more calories than berries. On the way back to camp I also grabbed some reindeer moss.

Already four in the afternoon. I was dying of hunger. While getting a drink, I managed to catch about fifty black and gooey tadpoles, whose eyes were almost as big as their bodies. I'd been observing them with disgust since yesterday, but I was finally ready. After squeezing out their intestines, I roasted them on a flattened rock and prepared myself psychologically to gobble them up. YEURK! Once they were shrivelled, I obtained the equivalent of two tablespoons of tarry guck that I pressed into two mounds. I closed my eyes and hesitantly gulped down the two huge pills. I would have been better off chewing on a piece of rubber!

The reindeer moss was too light to budge the scale's needle. On the other hand, there were 140 grams of Wild Sarsaparilla root. It had to be equivalent to five hundred or six hundred calories. I popped a tiny piece into my mouth and chewed with gusto. Hard as wood. No taste either. I chewed, chewed, and chewed. After five minutes I managed to swallow. I looked at the stack of roots. No way we could eat all that!

After the appetizer I served myself a portion of dessert — toasted reindeer moss. It was dry as a bone and tasted like raw flour mixed with dried mushrooms. There's a helluva difference between an edible plant and an eatable plant!

Because of the tadpoles that I gutted a while ago, an insane number of excited houseflies had joined the party around me. It made me think of Africa where these pests harass people on a daily basis. I ran away to the shoreline. Having sore feet, I decided to change socks. Right sock on left foot, left sock on right foot. Felt good!

Jacques had tried countless times to communicate with the CJMT radio station that day. Finally, success. They broadcasted our interview live, which boosted our morale. For the first time since we were there, the rising wind let us finally enjoy a brief moment of respite. I spoke too fast. A fly had flown into my eye and was crawling around in there! Jacques finally managed to remove it. I rubbed my lid with a dirty hand. Whew!

The wind died down again. I shuddered as a bizarre and waving purple sky blindfolded the reservoir. It became a true sea of oil, dark and mysterious. Suddenly the blackflies agitated themselves. They became absolutely furious! Indescribable! It seemed that it was raining on our heads, but it was flies bumping into our jacket hoods. We fled to the shelter, which seemed a thin protection in the face of the powerful and terrifying force of nature. I couldn't get to sleep. When would the storm unleash? To divert my sombre thoughts I worked on a birchbark berry basket. Soon fatigue gained on me and my eyes closed all by themselves.

When the storm came crashing in the middle of the night, half the shelter caved in. A log glanced off my nose, but the result was more fear than harm. We were too sleepy to worry about it. Scrunched under the remaining roof, we put off the repair until daylight.

It was 4:00 a.m. and I was frozen stiff. Cold scared me much more than flies. Bugs wouldn't kill me, but hypothermia could. As we became weaker, we might no longer be able to fight off the damp and penetrating daily onslaught. Nights required great courage. It was only the third; there were twenty-seven left. I was scared, deathly scared.

I woke to a thin frost that covered everything out of the fire's range. My breath turned into puffs of blurry mist. I sniffed all night. My nostrils were sensitive and my eyes felt like they'd swollen and merged into a single hole. I was afraid to face another day. Come on, André, don't be such a wimp! But even Jacques, who had thus far endured his ordeal without hardly a word, complained of rheumatic pain and discomfort.

The radio wouldn't connect again. What a trap we'd set for ourselves. Because of our means of communication and the publicity thus generated, we were stuck there for a full thirty-one days, and according to our own will it seemed! At least if we were really lost we could nurture the hope of getting out of there. But for the time being we were truly wedged tightly into this miserable adventure, mostly because of my big mouth. What a drag!

The welcome dry and warm wind let us fix the shelter in relative comfort. We solidified the structure, replaced the fallen roofing material, and chased down more. We also spent considerable time caulking the walls with peat moss to avoid drafts. Despite sore hands and bodies, we continued the back-breaking forced labour.

Since that morning I'd been concerned with Jacques's arthritic woes. I was convinced it would help if we could stay dry. I lifted our bough mattresses and felt underneath. The moss was completely soggy; we were lying on water! No wonder we were so chilly. I decided to remove our beds and scrape the entire sleeping surface until they were free of moss and only mineral soil remained. Then I spread embers and added branches to light a dozen tiny fires spread out all over the interior space. While the fires dried up the teepee floor, we carried tons of firewood to the shelter, enough to last several days.

Our teepee-shaped shelter.

I tried to chew another Sarsaparilla root, but it was so tough I couldn't even swallow it. I was fed up with survival books. Do this, do that, everything looks so easy. The obviously couch-potato authors vaguely explained all kinds of worthless and untried techniques. I swore that if ever I wrote a survival book of my own I would only describe skills I had experienced first-hand.

Ouch! A wasp stung me just behind the knee, and I'm somewhat allergic. I'd probably swell up a bit. Bah, I should stop whining like some old granny. I stooped to catch a *Bufo americanus*, a brownish toad with two poisonous glands on its back. I tied it onto my key hook and tossed it in the water. Maybe lady luck would smile at me.

I wasted the beautiful afternoon vegging out and sharpening another key into a better knife. I should have saved that kind of work for a rainy day, but I was simply out of steam for tougher tasks. By suppertime hunger forced me to charge into enemy territory again. As I advanced further than before, I was encouraged by the sight of a couple of berry bushes in the distance. But the blackflies wouldn't let me through. Who would win the battle? Like charcoaled commandos, they concealed themselves under the wet moss, organized their ambush, prepared for combat. I could not avoid the blast of their fire power and suffered the full force of their first assault. I advanced wavering and tripping, while hundreds of black aircraft took off to shoot me down and finish me off. But my tenacity surprised them; I pursued fruit pilfering more deeply into their lands. They deployed new recruits. After two hours of bloody war, weakened, I retreated. I sought refuge in my smoky fort to treat my wounds. The campaign count revealed a meagre five hundred grams of berries, which I shared with Jacques.

For the first time, I fell onto dry bedding with some semblance of satisfaction. Surely but slowly, I must have been adapting to the hostile environment. I even delighted at the sweetly setting sun.

In the wee hours I was sleeping lightly, suffering as usual from the cold and discomfort. Suddenly, Jacques shook me very hard.

"André-François, I think the fire is out!"

"You've got to be kidding. Can't be!"

I jumped to my knees and joined Jacques in stirring the ashes with a stick. Like panicked fools, we graduated from a careful investigation to a frantic dig. A small ember must be playing hide-and-seek somewhere in the cursed pile of white cinders! But we couldn't find it. I shoved my hand straight into the sooty residue and felt slight heat. Hopeful, I scratched and dug some more with my bare hands.

But in vain. The fire was dead! We were dead! Jacques couldn't encourage me; neither could I for that matter. It was impossible! I couldn't believe it, our friend Mr. Fire, dead! Just as things were looking up. Even with best intentions, I doubted whether we still had the minimum physical strength required to swing the friction bow. And as per Murphy's Law, for a fraction of a second bright lightning turned night into day, followed closely by ear-shattering thunder. The wicked squall approached, taking advantage of our weakness to attack. I was on the edge of tears. Would we shiver without fire through the storm before being able to call for help? Would we find the courage and energy to resurrect our best friend?

Emotionally terrorized, I stumbled out into the dark and moseyed on over to the lake. Between two fierce clouds a few remaining dazzling stars shone. Impressive it was, this flamboyant nature. I felt so infinitely insignificant. Another bolt of lightning split the sky in two, followed by a deafening clap. I thought of my family, my friends, Jean-Claude, the public. I couldn't believe that I would disappoint all of those people. What a rotten curse!

Ever so slowly I returned to the shelter, overwhelmed and shocked. I was already cold. It was a thousand times darker than usual. The glow of the shelter had disappeared! Our wonderful fire, gone! I swore I would spend every last ounce of my strength before surrendering.

Nineteen

Happy Birthday

> "The thornbush is the old obstacle in the road. It must catch fire if you want to go further."
> — Franz Kafka

I remember collapsing in the pitch-black shelter, dazed and confused. In silent denial, I just couldn't believe we had lost the fire. Restless, I exited and groped my way to the lake once more, passing by the blackened stump as usual. The stump! Maybe? No. A slim chance? I dropped to my hands and knees and started digging like a madman. Yes! The ground below felt slightly warm. I yelled to Jacques who helped me unearth the roots, at the end of which we hit upon a precious, tiny coal, and managed to blow it into flame just before the rain came. All the while we felt we were acting out the pre-arranged script of a cheap action movie, with our heartbeats as the soundtrack. And, as per the traditional last scenario, we danced, laughed, and screamed at our success. This close call taught us a valuable lesson and from then on we gingerly babied that fire, tenderly feeding it fungi as a precautionary measure. Silence resonated in our brains like the loudest of alarm clocks, whereas sparking wood lulled us to sleep.

In the days that followed, we gradually acclimated to our predicament in the infested boreal forest, like a mountaineer faced with thin air. Our meagre sustenance consisted of fruit picked from *Amelanchier* bushes, otherwise known as juneberries, serviceberries, or saskatoons. As we expanded our horizons, we found dispersed patches of these, at one point picking a full basket. Nevertheless, our average intake was less than four hundred calories per day, a long way from the usual three thousand back home.

After two weeks we were substantially weakened and fatigued by a weight loss of nearly five kilos each, and daily tasks demanded added concentration. Much of my time I spent like a tramp, vagabonding along the shoreline and begging for a few rich moments. Or peering out the tee-pee's triangular door into another dimension, that of drippy wet scenery, where lonely but lovely moments invited deep reflection on man's place in nature. How neat to marvel at the minute water meteorites crashing into the mud to form momentary craters. And then try to transpose that image to ponder life's vicissitudes.

Berries as appetizer, berries as main course, berries as desert.

Happy Birthday

On that occasion, the mini water holes I was observing served as portals into prehistory to reflect on how any human intervention affects nature. I imagined myself as the first man on Earth, in a pure, virgin forest. I was digging a well with my hands, a long and painful process. To help, I tore off a tree branch. This was the first impact, the very first use of nature. To speed up the process, I broke stones to make tools to cut a tree and transform it into a full-fledged but heavy wooden shovel.

But being ambitious, I wanted to go faster still. I fetched iron ore from the mountain. Then I erected a huge clay oven where I melted the metal into a light shovel, which permitted completing the well in no time. Bravo! But then I had nothing left to do. So I decided to carve totem poles to represent my greatness as inventor. *Voilà*, I had crossed the threshold into unnecessary impacts on nature, unexpectedly driving away the birds that once sat on the tree branches singing pretty melodies.

As I pursued my thinking to reflect on the repercussions of digging wells with gasoline tractor-drills it became quite clear that every produced item, be it need or luxury, destroys a corner of nature. Perhaps we should change the paradigm from production-no-matter-what to minimizing impact on nature.

I looked up at the moving sky for answers. With a bit of concentration, the clouds became stationary and it was me who was in motion, sitting on a planet. And Earth was like a boat sailing on a sea without end, filled with people of all races, religions, and superstitions. And if all these people sitting on this small vessel shared the one giant fish they catch per day, they would all be well fed.

Incomprehensible to me was the actual situation on our boat Earth. The stronger few keep the two fish filets for themselves and leave only the guts to the many others. This creates conflicts and as the enemies kill each other, the hull is pierced by stray bullet holes. Do they not realize that they risk sinking the boat? And that the only solution is to improve the lot of all, fairly?

As I lay there pensive, my mind wandered to the city: stress, noise, speed, pollution, drugs, cigarettes, power, possession, disease. I considered staying in the bush, away from that absurd society. The bark container full of clean water at my side comforted me more than would have a safe filled with money.

Over time we did manage to improve the teepee, slowly but surely, like a farmer sows his seeds. Once the fruit of our labours had matured, whether the rain violently whipped the shelter or softened and teased us with intermittent droplets, we gladly harvested its sweet-tasting security. Particularly appreciated also was the elephantine blanket I had painstakingly constructed from tall grasses dried in the sun, tied into bundles with spruce roots. As were the toque and extra socks I fabricated from the sleeves of my sweater.

Our main worry was keeping the fire under surveillance on windy nights, when we would wake covered with ashes. Twice our beds nearly caught fire.

Blackflies harassed us severely for two weeks before our toughened skin stopped reacting to their fading numbers. At one point I counted seventy-three bites on the back of one leg; I had not noticed the miniature entryway burned into my pants by a flying ember. When Jacques dropped a very rusty can he had found into my lap, I attempted to fight the bugs by dropping a smoky mushroom into the container and hanging it around my neck. The only result was an intense coughing fit. Imagining that someday nature would trust me enough let me in on her bug repellent secret soothed my agony and eased my bitching.

Many adventures punctuated what could have been a quite monotonous existence, mostly because of my inability to stand still. Indeed, my propensity to experiment non-stop soon got me into serious trouble. It all began seven days into the trip, when I started fantasizing out of boredom. Before me, the sea. Ah! If only I could sail freely! How handsome I would be with wind-swept-hair and waves whipping my face! Fishing, exploration, adventure! Mermaids! Why not build a ship? On my first attempt I entered the water up to the waist and using my remaining shoelace and belt, plus a few roots, I managed to tie together several logs. I hopped onto this raft that reminded me of my childhood hero. The proud and experienced Captain Haddock climbs to starboard and holds the helm. *Saperlipopette! Tonnere de brest!* This nut shell won't budge!

Happy Birthday

I ended up with the ridiculous raft a hundred metres further, soaked on the outside and drained on the inside. I sought refuge in the teepee, chagrined by my poor performance. But I hadn't exercised all of my options. *Mille millions de mille sabords!* A true captain wouldn't give up after having been shipwrecked but once! The next day, stubborn Haddock was at it again. What about a simple pirogue? With levers and small logs as fulcrum point, I managed to roll a huge tree trunk into the water, which fell in with a splash. I found another such trunk a hundred metres from the first and dragged it in the water until it lay parallel so I could tie them together.

But as I grabbed a pole and jumped on, the pirogue flipped over and I was tossed into the drink, head-over-heels. Twice. That sparked the idea to add an outrigger, which I solidified through triangulation with a diagonal pole. I was now master of my vessel.

With morning shadows, I took off my boots and tied them around my neck. Standing on my proa-style raft prototype, long slender pole in hand and flat root on deck, I set course for the nearby island to explore it and gain perspective of our territory. Splendid landscape. The raft split the inky calm water gently, silently. On the far side of the lake the elegant mountains proudly stood above the foggy water. What decor!

I wondered what I would discover there, across the lake. The vegetation appeared to be different, with larger trees. I reasoned that perhaps the south sun created a warm microclimate there. The pretty water and prospect of discovery became irresistible. So I sat and switched to J strokes with the rough improvised paddle and made my way across. It was slow going, but painless, until three quarters of the way. Then a light breeze rose, forcing me to paddle on the other side to maintain direction. When the breeze turned into wind, the strain caused a cramp in my left shoulder. Phew! Behind me, only a distant trickle of smoke indicated the position of the camp.

Staring into the deep and black water scared me. I suddenly became aware that I was infringing upon every safety rule in the book. No life jacket! Jacques didn't even know where I was, and even if he did, what could he do? Paddling faster, I finally docked at the rocky point's natural wharf, completely exhausted. Where was I to find the energy to go back?

On shore I found a few raspberries, which I gobbled down instantly. Then, to make a long story short, I wasted my time hiking part way up the mountain and looking around, worried all the while I would get lost. As minor positives, I did find a cherry sapling, which I broke off to use as a future hunting bow, and gathered an elderberry branch with the intention of hollowing it out to make a bellows for the fire. But my thoughts were entirely consumed by the scary ride back. And scary it was indeed.

The wind had increased in velocity and the waters were hardly reassuring. Knowing the situation could only grow worse, I jumped onto my craft and the torment began. I had paddled three hundred metres or so before I realized something was amiss. My boots! I had forgotten my boots on the shore behind me!

It took over an hour to fight my way back against the wind to retrieve my precious footwear and get on my way again. The wind was shoving me diagonally, deviating me significantly from the path. So I had to ask my too-short root paddle to maintain a forty-five degree angle to the waves.

Near the halfway point, again as in a movie thriller, the intensity of the wind increased and waves were now washing right over the raft. I was barely progressing, like a snail across a highway. To be blunt, I was scared shitless. And that was before one of the lashings broke and my outrigger started falling apart. I jumped in and used my handkerchief to tie it back up. I managed to roll back aboard and paddled like hell strictly on adrenalin until I crashed aground. Land! To complete the cinematic display I kissed the shore and lay down, dazed, soaked, and exhausted. That night, the cold's claws seemed sharper than ever.

Other highlights of the first couple of weeks were entirely related to the unrelenting search for food. We had observed a couple of red squirrels frolicking about fairly close to the campsite. Jacques advised that I would be wasting my time setting up a squirrel pole snare like those described in survival books. Naturally, like a kid contravening his parents' rules, I set one up anyway, "borrowing" a piece of wire that served to attach the radio's handle. But experience proved my companion right, as I observed the squirrel avoid the snare each time he climbed the path.

But a couple of days later, as I was moseying around the next bay, a squirrel attracted my attention and I froze, observing it. Hmm, interesting. It hopped onto a floating jumble of logs and climbed a rotten tree that stood erect in deep water some five metres from shore. It entered a knotty hole, which I presumed was its nest. Picking up a long pole to serve as a lance, I approached by balancing upon one of the log bridges that led to it. The other log paths I pushed away, reducing the squirrel's access to shore to the single one I was standing on. I positioned the tip of my spear at the edge of the hole ready for a fatal strike and kicked the tree. But in the wink of an eye the lighting-fast animal had leaped out and scurried between my legs into the forest.

Curious that I might find edible nuts, I started shredding the half rotten tree trunk with my spear, when I was surprised by acute and plaintive cries. Baby squirrels! With shrieks having alerted the mother, she came scurrying back to the rescue, passing by me somehow and snatching a pup in her mouth to carry it off into the forest. Watching my supper pass before my eyes like that wouldn't do, so I ran to get my squirrel pole trap and laid it against the edge of the nest. I hit the tree again and the whining pups called to mom again. Too hungry to dwell on the philosophical issues haunting me, I readied my spear. Unbelievably, the nimble squirrel managed to avoid both my strikes and the snare while reaching her offspring, but got caught in the snare on its panicked way out.

Even the most seasoned trappers acknowledge that seeing an animal get caught in the trap is not quite like gathering it up once already dead. Truly, I wasn't too pleased with myself. The lean meal I would partake in that night did not compensate in the least for my feeling of guilt at having removed this beautiful beast, mother of needy children, part of this nature I was striving to protect. On the other hand, it was a unique occasion to reflect on the difference between killing to obtain meat or paying others to do so. It is difficult to associate "McCroquettes" with living beings. That, is the duplicity of life in the city. There in the wilderness reality jumped right in my face as I spent a night listening to the plaintive cries of the five pups huddled in my hat, and the next day too, as the wicked and monstrous hunter forced himself to cook them up to respect the survival mantra that states "you kill it you eat it." Having weighed my meal, I concluded that this must have been the strangest "quarter pounder" ever. Poor Jacques had to leave the scene. Like he had the previous day when I had prepared "Squirrel tripe stew à la rusty can." As I sat alone in the restaurant, I too had complained. Maître d'! There's disgusting white foam in my bowl, and the broth tastes like dishwater!

But switching dinners didn't help. My meal of whole minnows boiled in the same can didn't taste much better. Not even fit for cat food. I had spent an entire day lugging logs and rocks to fence off an area to trap the million of small fry I had noticed darting about in a shallow spot near shore. Having traded hats with Jacques, I used his net-like baseball cap to chase them down. End result before they scattered: a piddly 175 grams. The weather never permitted me to retry.

Even on the sunny days when it was possible to fill my tank with berry fuel, my motor didn't run well. I needed some premium gasoline, otherwise my carcass would end up at the scrap dealer.

On day six I was combing the shoreline to the east for I-don't-know-whats, when once again I heard the loud splash of what I could only figure was a bullfrog diving and admired the pretty pattern of concentric waves. Further on, the phenomenon repeated itself. I imagined another pair of powerful and delicious legs propelling a green slimy body. If I walked with feathered steps, could I surprise a third? Oh! An inanimate object was lying parallel to a floating log. A filiform pike! Another step and splash! It rushed to deep water, striking its tail. The mystery of the invisible frog was solved. Sun-tanning fish!

If only I had a decent hook. Wait a minute! What about the thin metal spring inside my anorak's cord lock? I ripped it out and stretched it, then broke it into three equal pieces by striking it repeatedly between two sharp stones. Using twisted strands of thread from the inside of a piece of worn out lace, I bent the three sharpened points and tied them together into a treble hook. A folded ten-cent piece added weight to my creation. After several hours of patient twisting of the same threads into a line, I was the proud owner of a brand new four metre long fishing line. I hung a few tadpoles on the hook and roared over to the bay to launch the bait.

I returned to camp empty-handed and shared my findings with Jacques, who was not impressed whatsoever. The previous day, he had seen five such pike sunning themselves and had even managed to touch one with a pole. His patience impressed me. But how to catch them?

Happy Birthday

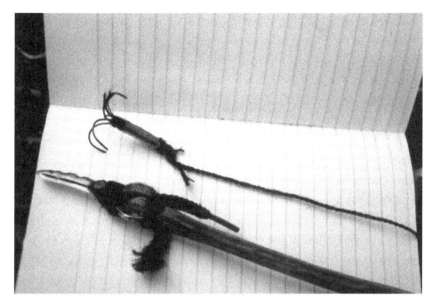

A knife improvised from a key. The treble fish hook was manufactured from a cord lock spring and a folded dime.

If Jacques touched one, I figured I could certainly manoeuvre my hook under one. I returned to the bay just a little too abruptly. Splash! I had to show way more patience just to get a glimpse at one of those elusive pikes. So I slowed down. Way down. I let my feet slip gently, lightly groping the soil. Moving no more than a centimetre at a time, with extreme delicacy, I introduced the grapple in the water two metres from the fish. More gingerly still, I nudged the hook toward the target. Evil blackflies were biting me around the eyes and one was now crawling up my nose, but I dared not move. Five minutes passed. Nice work! The treble hook was now just below the fish, the line touching its side. Jerk!

The fish was propelled into the air. But the hook wasn't sharp enough to pierce its skin, the tip had simply bent out of shape. Damn, damn, and triple damn!

I wandered for three-quarters of an hour scrutinizing the logs before seeing another pike. Same scenario, same result. Then a third failure. All of these immobile fish, within our reach, without being able to capture a single one! I swore that after the trip I would always carry a hook in my wallet.

I continued dreaming about fish. Perhaps I could weave a net to span the three-metre-wide gap between the tiny island and shore? After all, Native people used to do this, using basswood, dogbane, or milkweed fibres. After spending countless hours weaving string from willow bark, I attempted to concretize the project. By nightfall, out of frustration, I tossed the twenty centimetre fishing-net toy deep into the bushes.

The next day, Jacques came into camp relating excitedly that he had almost caught a fish. He had managed to use the partridge catching technique to partially slip a shoestring noose over the fish, but missed because the lace floated out of shape when submerged. I suggested he use the wire from the squirrel snare. That would be cheating, he replied. True, I answered, but just try it to see if it can be done. His thrilled yell that afternoon left no doubt as to his success. Pike! He explained that the trick was to pass the noose head first, contrary to our previous tries, and to pull as soon as it reached the gills, at which point the fish darted forward so quickly he was caught near the tail.

I really enjoyed that quarter pounder of fast food too, all by myself. Picky-eater-Jacques disdained fish and no amount of coercion could convince him to replenish his protein supply. Again I laughed at popular survival literature that suggests eating larvae and insects. No way! Unless at death's door, no one in our contemporary society will eat such fare. I for one wouldn't even consider eating earthworms until having sunk way deeper than I was thus far.

Following Jacques's example, I fished every spare moment of that entire week. After fifty hours logged at the task, I had managed to harvest only two more fish of similar dimensions. Then the weather patterns changed and the floating fish disappeared into the depths.

On day thirteen, Jacques asked me to listen as he pushed the playback button on his tape recorder. I immediately recognized a beaver-tail slap. On the small pond half a kilometre east of our camp, a yearling had begun building himself a home, the telltale sign being the tiny dam he had commenced working on. Jacques swore we would be eating juicy cubes of beaver steak before his birthday, on day eighteen, five days from now.

Fast-forward to August 18th, in the wee morning hours, just before sunrise.

Happy Birthday

"FIRE! FIRE! WAKE UP YOUR BLANKET'S ON FIRE!"

I leapt up, smashing my head against the sloping roof of the shelter. UGH! What the hell? What was this intense light? The heat! The smoke! Jacques was shouting, madly smacking at my grass blanket with a stick. No! The fire jumped to the birchbark ceiling. The shelter was on fire! NO! We were surrounded by flames! My god!

"GOTTA GET OUT OF HERE!"

We dove out of the rising inferno, rolling on the ground to extinguish the sparks on our clothes. Instinctively I grabbed the corner of the fiery grass blanket and dragged it outside and managed to smother it against the damp earth. Crap! I'm missing a boot! Desperately, I plunged inside and snatched the precious item, to exit once again with a judo roll.

Dazzling flames were already soaring several metres into the air. My glasses! I needed my glasses! I tried to leap into the hellhole a second time, but the fire was too intense. Maybe I can rip through the wall of the shelter from the outside to reach the spot where I had hung them last night. Quick as a rabbit and with the fierceness of a wolverine I slashed and gashed at the grass and bark structure. YIKES! Our recorders are in there too! And the camera! Jacques helped me in a frantic but futile attempt to pierce the fortress. The jostling heat forced us to retreat several paces.

We were appalled by the power unleashed before us. The scorching tower rose to the height of the largest trees surrounding us. They trembled violently, threatening to ignite. No! Please, no! Not a forest fire!

The shelter crumbled. Phew! We could finally approach and hook a few long logs to pull them out of the fire. Little by little, we took control. The danger had passed.

We were still in shock. Jacques ran to the radio:

"MAYDAY, MAYDAY, this is an emergency call. Mayday, does anyone read me? All stations, all stations, Mayday, this is an emergency call, please answer, please answer. Damn! No one hears us!"

Motionless and silent, we contemplated the giant red circle that had closed in on our treasures. My beautiful basket overflowing with fruit! My little green sleeve-socks! The cameras, the voice recorders, our friction fire kit, the other baskets, our homemade ropes, the weigh scale, the thermometer, Jacques's t-shirt, my handkerchief, my eyeglasses, our key-knives, our belt-buckle axes … everything, everything, everything was gone!

5:15 a.m. I didn't yet fully realize the consequences of the fire. Must our journey end thus? I was trying to analyze the situation. Could we continue? Jacques was obviously hooked on leaving. He continued to call for help, over and over and over again, but there was no response. Security my eye, this radio!

A lazy stroll along the shoreline will do me good. A smiling log invited me to sit. I drank and thought. At least my scientific data was spared! Good thing we had taken the habit to store the 35mm film canisters and the voice recorder tapes with the radio in the waterproof bag.

I compared our situation with that of the first day. At least we had fire. Too much of it that morning, mind you. Flies were less voracious. We were on a fine site, despite the fact that it was now rather depressing. We'd toughened up. On the other hand, we'd much weakened since the beginning of the trip.

Perhaps the media would perceive this misadventure as newsworthy, which would help Jean-Claude promote Nature's University. In a sense, maybe the fire wasn't so bad after all. The day before I was bored to death; now I faced a good challenge! Okay, it was decided. I was staying. At least I could endure a few more days.

Jacques was still trying to establish communication. It was his birthday, poor devil. He must have thought the candle was a tad big! He looked so depressed! I suggested a diversion.

"Say Jacques, why don't we go check out your beaver trap?"

"Okay. This cursed radio won't work anyway."

"Tough on the psyche, eh?"

"No kidding! Let's not talk about it."

We clambered up to the pond. The rudimentary deadfall trap hadn't budged, and therefore neither did Jacques's morale. It must have been frustrating beyond belief for him to have failed capturing the beaver after five days of attempts. Alas! As I suspected, it wasn't so easy to substitute wooden deadfalls in lieu of standard metal traps.

I didn't know what to do to cheer him. Personally, the fire incident didn't bother me anymore, I had already accepted it. In fact, I felt as strong as a bear. The only thing that annoyed me was not seeing clearly. If only I could have saved my glasses!

We returned to the desolate campground scene and like anthropologic scavengers we started poking around in the fire. We discovered the remains of a weigh scale base, the skeleton of a Dictaphone, and a piece of plastic melted around a camera lens. A camera lens? Wait a minute, didn't the glass burn? That meant my eyeglass lenses lay somewhere beneath these ashes! We sifted through the hot cinders with sticks and found one, then the other. But their surface was completely charred — they were opaque and useless.

I began making plans for the new shelter. Because of the cold end-of-August weather we were expecting at this latitude, we decided it should be much smaller with a fire in front; the best shape would be a triangular lean-to. When the fire died down, we would scrape the remaining embers aside and build the shelter in the same location. At least the ground would be dry.

For the time being, there was nothing to be accomplished in that sense, so off we went berry picking. This was the first time we went together, since for once there was no danger of losing the fire. That precious fire that until then had provided us comfort, warmth, and security had now literally turned against us. Why? I suppose that white can't be fully appreciated without black to compare. It's nature's way. A good excuse to not blame myself, obviously.

I finally mustered up the courage to engage in conversation with Jacques. I agreed with him that it was tempting to use our misfortune as an elegant doorway out of there. But as I exposed my plan for the new shelter, I managed to convince him that our fate was no worse than it was the day we arrived. I also insisted on the added hype the event may have on the promotion of Nature's University.

"In any case, Jacques, I think we can at least build the shelter and survive until tomorrow."

"That's what's sickening. I know we can get by until tomorrow without a problem."

"So?"

"So, damn it, I know that if we make it through tonight, we'll have to stay until the end of the month."

We broke out laughing. How funny, we were reading the exact same script. Much worse than our sinister incident was being trapped there for an additional thirteen full days! And as our nerves calmed down, we could truly feel how our rude awakening at 5:03 that morning had added to our cumulated fatigue.

With the rise in barometer we eventually established a communication with Jean-Claude and dumped news of our disaster on him to see how he reacted.

"It's obviously impossible for you guys to stay out there in those conditions. I'll send you a helicopter as soon as possible."

"Very nice of you, Jean-Claude, but we've decided to stay. We like it too much here!"

"Huh? You mean to say you want to continue the adventure anyway?"

"Yes, Jean-Claude, we believe too much in the cause. We'll try to stay a few days more. Do you think it'll help pass your message?"

"Count on it. I find you guys extremely courageous. You two are aces! Good luck!"

Flattering. How I love surmounting the impossible! To me there is nothing in the world like a good solid challenge. Of course I have to pay the consequences for my sin. What is clear, in any case, is that if I were forced to choose between feeling pain or feeling nothing, I would choose the pain without hesitation. At least suffering shows me that I'm alive.

But Jacques didn't fare so well; he became doubly depressed. During our radio interview, he could no longer control himself and broke down, sobbing. I had to replace him at the mike. It wasn't easy.

We began the reconstruction of the shelter. Log by log, the new structure rose. Fortunately it wouldn't rain that evening, for bark was now rare in that neck of the woods. We built a temporary roof using pieces of wood caulked with moss. It wasn't much protection, but it would have to do for the night.

We still had to carry wood for the hours of darkness, but I needed a pause really badly. I headed to the beach to recline on my favourite log. The waves on the lake looked like white lambs dancing on a blue pasture. For the first time, the shepherd wind led the herd to the southwest. Does

Happy Birthday

this indicate a major weather shift? How will we fare in our tiny shelter when furious lightning comes to whip the sheep and thunder barks loudly at them?

I took my boots off to rest my feet. I imagined myself barefoot looking for shoes. A pair of sandals woven with bark like those I used in Colorado wouldn't protect me from these flies. No, better catch a beaver to make leather moccasins. What a job! First, he needed to fall in the trap, Mr. Beaver. Second, we needed to skin it and scrape off the fat, all with primitive tools. Next would come the tanning process using brains, fat, and ashes. Only then could we finally cut pattern pieces and use a bone needle to sew a pair of rough moccasins.

Then I imagined myself in the city, without shoes, and without a penny. After working two hours to check some vending machines looking for forgotten coins, or by picking up bottles discarded by naive or reckless polluters on the side of the road, I would quickly obtain a dollar. A few minutes later, I would leave the flea market, proud owner of old-fashioned but sturdy shoes that would greatly surpass the comfort of the moccasins made in the forest after several days of intensive work.

A hundred metres along the shore lay a few huge rocks we would need. Using the partly demolished raft, I floated them back to camp. These I rolled over to the fire pit. During the night I would use some forked sticks to tumble them over to the inside of the shelter to warm our feet. In spite of our minds' command to build a bed, fatigue pushed our stop button. At least we still possessed my grass blanket to use as a mattress. It was almost as wide as our tiny shelter floor.

As Mother Nature's maternal darkness started tucking in the day, I mustered up the courage to go accomplish a final task. An idea had been trotting in my head all day. Using the sharpened key I'd recovered from the ashes, I cut a small plate out of birchbark. On it, I installed a small ring of the same material held in place with a toothpick. I grabbed a few berries and crushed them into a paste to fill the ring. This would serve as the base layer of the "cake." On top I spread a glaze of delicate white wintergreen flowers. The grand chef was about to create a masterpiece! A few decorations with some red bunchberries and heart-shaped wood sorrel leaves and the cake was finished. Cute! I stuck a miniature bark torch on top and approached the shelter slowly. Jacques looked exhausted, drained, completely demoralized. So I cleared my throat, lit the "candle," and sang a lively "Happy Birthday."

Twenty

The Philosopher

"Wonder is the feeling of the philosopher, and philosophy begins in wonder."
— Plato

"Hey Jacques, did you see Fatso? He's got a real blue stomach now. And Siamese is blushing."
"What about Crooked Face and Tiny?"
"They're between red and blue."
The tame blueberry friends perched on a lone bush right next to our shelter had become our "ripening index." We had agreed right from the start not to touch these four guys. They would enjoy basking in the day's sun.
I miss the old shelter! Crammed into our doghouse we slept, well, like dogs. The roof being barely a metre high, we couldn't even sit comfortably. It wouldn't be fun during rain. Speaking of rain, mares' tails indicated we might get some. Back to work! We thickened the roof with another layer of logs and spongy moss. And one more on top of that.
The night before we burned the scorched remains of our former shelter. The rotten and green logs smoked us like sausages. We needed a new provision of dry firewood. But true hunger sent us berry picking first. The juneberries were gradually withering away, whereas the blueberries

Jacques sitting in our "doghouse."

were ripening quickly. I'd never before witnessed this transition, which seemed made-to-order to sustain life. But pickings were slim; we seemed to have cleaned the juneberry bushes from the entire flank of the mountain, from both sides of the swamp, and all along the lake. So I had to patiently harvest lone blueberries, one here, one over there. I returned to camp, having barely hidden from view the bottom of my hat.

At a pulse rate of thirty-nine beats per minute, motivation was rare. As I exited the shelter, I felt faint and wiped out, landing flat on my stomach. I dusted off my clothes, which were horribly dirty and worn out. All of me was horribly dirty and worn out, in truth. I tried to nap, but housefly pests prevented shut-eye with their incessant landing on my skin with a buzz.

I stared with melancholy at the two blackened lenses of my glasses. I tried to rub them with my shirttail, with a piece of wood, with ash. No change. There must be a way to polish them. Wait a minute, I know a plant that contains silica, in the genus *Equisetum*, that has abrasive properties. It's even called "Scouring Rush," because campers use it to scrub pots. Seemed to me I saw some on the other side of the island where the soil was sandy.

Once there, I grabbed a wad of the unique shoots and started rubbing hard one of my lenses, concentrating on a specific dime-sized area on both sides of the surface. After several minutes the results were encouraging, so I scrubbed some more. Fifteen minutes later I started seeing through!

Rub-a-dub scrub scrub. It's always the same in survival. When you aren't rubbing the point of a fire-by-friction spindle, a key, a hook, or a belt buckle, it's glasses. Fortunately, the Scouring Rush plant was not endangered. In his madness to gain control and material wealth, man had already managed to annihilate a considerable number of plant species, many of which no doubt contained unique and special properties. Did we unwittingly wipe out the miracle cure for cancer? Who knows. Yet, in all corners of the world, under the pretext of "progress," humanity irreversibly invades the last natural territories of our planet and continues to reduce our future options. A bit of wisdom perhaps?

Rub, rub, rub some more. That was somewhat better. I returned to camp wearing my new monocle, a pile of scouring rush in my pockets. I replaced Jacques as fire picket and started polishing the second lens.

Running out of patience, I tried instead to manufacture wooden frames, fearing never to achieve a satisfactory result. On one of the lenses, part of the metal frame remained to help me fix it to the wood. On the other there was nothing. I carved two arched nicks in a willow branch into which I inserted the top of the lenses. Then I carved a nosepiece. Using a short piece of the fishing line that still lingered outside, I attached all of the pieces together the best I could and wrapped a length of shoelace around my head, tightened by my remaining cord lock. Whoa! Quite the spectacles! Maybe I'd start a new fad: philosopher glasses.

Crooked Face and Fatso were looking at me funny, wondering what I was thinking. Well guys, one day while I was in the African village of Chikal, an individual approached and pointed to my glasses while uttering words that resembled a question. I let him try them on. His face was instantly illuminated, as he lifted his arms in the air and began running

Philosopher eyeglasses.

and screaming like a madman, seeming to say "I can see, I can see, it's a miracle!" Then he realized that his happiness was short-lived, as he had to give them back. Fortunately, I had an emergency pair to donate. Probably the prescription wasn't correct, but he sure was pleased.

It boggles my mind to think of all the people on this planet who still don't have access to basic medical care or the slightest correction of their disabilities. There in the forest, we at least had the assurance of our eventual return to pampered living. Without a shadow of a doubt, that made me better off than all of the world's poor who were stuck in their misery their whole lives.

My improvised glasses gave me a new outlook on life, and it compelled me to philosophize some more. My lengthy stay in the wild seemed to have purified my soul, and I felt I had become but a spokesperson to preach nature's teachings.

What use could we make here of a padlock, escargot dishes, champagne flutes, a dustpan, an electric toothbrush, or an ironing board?

Ah! The incredible ease of modern life with its benefits and advantages. Mankind has come a long way since primitive living conditions. Medicine, clean and heated shelters, diverse and abundant food....

If only humans had wanted to stop there. But no! With lack of foresight, they have pushed the limits of comfort until glut has replaced it, at the edge of the absurd, and they now suffer the repercussions of this excess in the form of abominable pollution of air and streams, stress, numbing walls of concrete, drugs, and suicide.

Hey, feels good to criticize. High heels! Society suffers from the high heels syndrome! With the ability to wear soft shoes more comfortable than the Neanderthal could imagine, society artificially coerces people to wear footwear so poorly adapted, so harmful to the feet, that if I had some here I would prefer going barefoot. Improvement? No, nonsense! What comfort can we find sitting in a straight chair all day long in stuffy buildings which re-circulate pollution, wedged into a jacket and tie, deafened by engine rumbles, drab asphalt, and artificial discussions? Is a lion in a cage still a real lion?

How much are calm, serenity, simplicity, inner peace, sincere friendship, and tenderness worth? A simple look at nature suggests that there is a definite link between our extravagances and the fact that Third World countries stay third. And this, in spite of the knowledge that the planet has enough resources to bring each and every person to a truly comfortable standard of living.

I spent the rest of the day in my philosopher eyewear, dwelling on the solutions to these puzzles. I made some half-hearted attempts at shaving down my cherry bow stave, replacing the handle on the key knife, and gathering wood. Laziness sang me a lullaby and cradled me into my camp bed. I asked to go pick berries, but she kissed me again. I could not resist her charm.

Starving as I was, the sky became blue frosting on a rich cake, covered with clouds of whipped cream. The sun poked through to decorate it with lemon-yellow flowers.

"So Jacques, what would you feel like eating for supper tonight?"

"I don't know, maybe roast chicken."

"Me, I would go for a wonderful '*Suprème de volaille cordon bleu*.'"

"What's that?"

"It's fantastically good. You take fresh chicken breasts that you split open and flatten, then stuff them with black forest ham and Swiss emmenthal cheese. You continue by rolling them in flour seasoned with secret spices, soak them in beaten egg and toss them in bread crumbs. Finally you seize the suprèmes in oil and bake them slowly at 275° Fahrenheit. The cheese melts inside and the chicken is so tender it melts in your mouth."

"Yum! What do you serve it with?"

"Usually with rice pilaf and a buttery green vegetable. But it could go with duchesse potatoes too."

"I would order strawberry shortcake for dessert."

"Ooh, that would be so good!"

"I'll eat three when we get back to civilization."

"Me too. Plus eleven chocolate donuts."

"Why not twelve? That would be a full dozen."

"Hey, do you think I'm a pig?"

Jacques chuckled as he pulled out his notebook to scribble recipes and techniques, as usual. Until then I had taught him everything I remembered from the chef's course concerning soups and sauces. We started covering main dishes, like roast beef au jus. My mouth watered while describing my specialties. But Jacques just loved those cooking lessons, so I put up with the subtle mental torture.

It reminded me of my dear papa with his tall white hat and unmatched enthusiasm, preparing tournedos, *paté de foie gras*, *tourtières*, and chocolate bombe Alaska. In the middle of this imaginary feast, I crossed the kitchen threshold and tumbled into the land of dreams.

What? I would swear that a noble Indian chief was begging me to follow him. Masked by war paintings and wearing a long feathered hat, he scurried mysteriously along a trail behind our camp. Curiosity prevailed and I followed him for hours, without ever quite catching up. Suddenly the forest thinned, and I saw houses, then a complete village. The chief told me to wait. Meanwhile, a few individuals recognized me and showered me with questions. I tried to explain, but they could not fathom why I was not in the forest. Soon reporters arrived, shouting. I wanted to run away but it was too late. They accused me of having invented a Survivathon scam. The scandal made headlines, people pointed and laughed at me. A crowd rioted. They wanted my skin! They caught me and tied me to a post. They would burn me alive! The flames rose around me. Yikes!

I woke up with a start to a too-hot fire. Phew! What a horrible nightmare, no doubt a reflection of my actual concerns. The door to this penitentiary only opens with a misfortune. And even if we try to escape, the ridiculous awaits. We were therefore left with no option but to suffer our entire thirty-one days of forced labour, prisoners of our solitude.

I slept pretty well, surprisingly. At home, not having taken a shower before going to bed, a lone buzzing fly, slightly damp sheets, a hard mattress, a hollow belly, a single mosquito lurking around, or even a wrongly set thermostat — only one of these annoyances would prevent me from sleeping comfortably. In our shelter, all those conditions were pooled and more. Despite all this, that night I was so exhausted that neither smoke, dirt, flies, branches poking my back, cold, moisture, nor hunger could disturb my sleep. The human body's capacity to resist and adapt surprises me more and more each day.

That day rain took on a different meaning, since I got to witness first hand its role in nature. Its regular rhythm hummed a melody. I listened with serenity to the first movement of the downpour symphony. We were once more locked into our dog-sized dungeon with chains at our feet, without even dry bread. There was absolutely nothing to do but wait, wait, and wait some more. I was going nuts! For our emotional balance, we obviously had to keep a bit of food in the hut.

The rainy symphony began its andante movement. The repetitive and monotonous theme wanted to lull me back to sleep, but I continued scratching at my bow stave.

The third movement, fortissimo, impressed with its intensity. With no food in sight I was fed with frustration and boredom instead. I wasted the day alternating between poking the fire, shaping the bow, and filling pages of my journal. Morbid thoughts filled my mind. Between the toes, we suffered painful cracks caused by moisture and lack of hygiene. I hoped we wouldn't end up with dreaded trench foot! It had been my worst day so far. I could eat anything right now, old dry bread, plate leftovers in restaurants, even fast food. Lingering until nighttime seemed interminable. Feeling sorry for myself, I didn't have to count many sheep before passing out.

By lunchtime the next day I couldn't take it anymore. If it rained all week, what would I do? Sit in the shelter without eating and wait to die? The drops tapped on the shelter roof as if to telegraph the answer. Damn it! I'm not made of candy, I wouldn't melt! Come on Bourbeau, get your ass moving! Are you a wimp?

I crashed into the forest like a bulldozer, mowing down the shrubs in my way. In seconds I was soaked to the bone; I might as well have jumped in the lake. The wooden frames of my eyeglasses fell apart under the pressure of my hood. Enraged, fuming mad even, I crossed the hill behind the camp with determination, circled the beaver pond, climbed over the mountain, and went down the other side, way beyond where I'd yet been. There I finally encountered some well-loaded juneberry shrubs that hadn't been touched. I manhandled the fruit, plucking them off roughly, in a daze.

Little by little nature worked its magic on me. With the forest refreshed by the downpour, everything shimmered and glittered of deep rich colours as if the hand of heaven had polished the land. Water beaded peacefully on the leaves. The dew glistened on fruit like diamonds. Silence. All around me was so beautiful, so peaceful.

I am ashamed. This morning I complained of my fate, I found life hard. I had not eaten, but I was sitting on my butt. What did I expect? That my lunch would fall from the sky? I peered at my hat full of food. Security? I have it in me.

The slope leading back to camp seemed easy now. From the top of the mountain, I admired the most fantastic rainbow I'd ever seen. As if a gift had been sent to comfort and delight me since I'd graduated with my diploma in wisdom.

I'd been gone for four solid hours. I must have gathered nearly three kilos of juneberry pearls plus a few glistening *Suillus* mushrooms. Cold numbed me, even while quickly hiking back. What a joy it was to take refuge in our warm and homey corner of the forest!

"Hi, Fatso!"

"How are you today, André-François?"

"It's not going so bad, thanks. I just got back from work."

"Always at the same job?"

"Yup, still a berry picker."

"Ah good! Are you well paid at least? Is it hard work?"

"Well you see, my dear friend, it depends on nature. Today was my best day so far. I picked about three kilos of berries in four hours, so I made a few bucks. But last week I picked for two hours and only made fifty cents. Not much compensation for a sore back, frozen feet and getting eaten alive by blackflies."

"Those are ridiculous conditions! You should complain!"

"That's life in this land. No unemployment insurance, no unions. So you do what you can. Remember the good old days when I earned more than twenty-five dollars an hour? That was the good life! To say that I dared to complain anyway!"

"Yep, I remember. Maybe you should have appreciated it more."

"You sure are right. Well, I have to go to bed now. Goodnight, Fatso!"

Late the next morning I was trying to tame the beast called solitude as I hovered my boots over the fire after another semi-successful run at picking berries. To dry them faster, I scorched a few small stones that I dropped inside and shook them about until they cooled. Jacques entered my field of view. He was strolling casually, whistling, with a grin spanning from ear to ear. And with reason. A giant sucker fish dangled from the snare wire!

"How in the world did you catch that?"

"It was easy. It was floating by the edge of the water, already dead. The wind pushed it ashore."

"Ah, okay. I was beginning to think you were a miracle worker. I wonder if it's good to eat."

"Are you crazy? I wouldn't touch it if I were you! Who knows how long it's been rotting."

"Well if it doesn't smell bad, I don't see why it wouldn't be good. Give it to me."

"You're crazier than I thought!"

I picked up the burbot. It wasn't pretty, with its ugly sucking mouth. But I didn't detect any odour. Cold water had surely kept it fresh. But how did it die? On his belly, a ten-centimetre gash gave me an indication. Probably it was attacked by a pike or a bird of prey. Unless it got hit by a boat propeller.

Maybe he died of some type of toxin? I wouldn't think so, because it was the only dead fish we'd found since we'd been there. Besides, the water wasn't polluted. Should be no danger if it was well cooked! But what if I fell sick of food poisoning? Bah! Stay in bed to wait or stay in bed sick with cramps, it wouldn't make a huge difference. If I was ill for a couple of days, I'd look forward to being better and time would pass by faster. The trip was nearing its end. My basket was full, the shelter was solid, and Jacques could maintain the fire. If I took a chance, I would have extra energy for a couple of days and I'd be able to get some work done.

I tossed the fish on the embers and waited until it was well done. I tasted it. Not as good as pike, but it was food! And it was big, surely a kilo. I ate slowly at first, then more and more avidly until I surprised myself stuffing my face. Soon I couldn't gulp down another bite. My stomach was knotted and hard as a rock. Why didn't I control myself? I'd burst like a balloon!

The Philosopher

The most interesting part of the burbot having being swallowed, the remains were less than appetizing. All of sudden I regretted my choice. I threw the carcass away in disgust.

I headed over to my straw bedstead and closed my eyes. The sucking, hanging lips of the bottom feeder haunted my thoughts, preventing me from sleeping. I felt like I was going to be sick.

By midnight, eyes wide open, my belly was just one big cramp. What a foolish thing to do, partaking in such a substantial meal after having starved for over three weeks! It turned out I was ill all night because of overstuffing myself, not from poisoning.

The next few days of the Survivathon found me seriously battling loneliness and boredom. Surprisingly, those two unsuspecting enemies left me wounded with more excruciating pain than I had thus far experienced. Perhaps other individuals would have profited from the forced retreat, but certainly not I, especially during this go-all-out-until-you-drop stage of my life. Progressively weakening from continued hunger (my pulse rate dropped to a record-low thirty-four beats per minute), I could no longer hide behind experimentation and was thus forced into soul-searching. And I didn't particularly like the self-centered reflection I saw in nature's mirror.

To escape the monotony that was invading every pore of my body, I spent much time writing philosophical essays sparked by observations of the moment. For example, when I finally captured my only Spruce Grouse of the trip and was plucking and cleaning it, I tore open its crop, curious to find what it was eating: some leaves and buds, but mostly blueberries. Fatso! Siamese! Crooked Face! No, it wasn't them, but others like them. I was startled by witnessing firsthand the obvious link, from sun to berries to partridge to me. The sun was feeding me directly. A much more concrete exposure to the food chain than that taught to kids in school. From there the step is slight to comprehending that solar energy supports all life on earth. Even oil, from which we manufacture plastics, nylon, and fuel, comes indirectly from the star we orbit around. It is simply organic waste, compressed for many thousands of years. But on the human scale of time, petroleum products are finite and not renewable. Thus the need to stop plundering them and switch our focus to other forms of energy.

That bird had provided the very best meal of my entire existence. It also gave me feathers for my arrows and wonderful pocket warmers. The next morning, I woke to an unspeakable appreciation for life. I called it my thousand-thank-you day. Thanks to nature for letting me live, thanks for warm sunshine, for fresh air, for clean water. Thanks for all these gorgeous blueberries. Thanks for this one in particular. And this one also. And this one.

Oh! How I appreciated the blueberries that day. Each one that ended up in my hat deserved a pause of acknowledgment. No comparison with former life scrambled my mind. I was locked into the here-and-now. The shelter appeared as the most charming of homes, and the fire a marvel. While rekindling it, my heart filled with joy. Sitting on my fresh grass mat, I savoured my blueberries one by one. Their texture tickled my tongue. They had grown before my eyes; they were my friends.

I stared at my boots with gratitude, imagining myself barefoot. And this shirt, these pants, this anorak, what a chance to own them! Around me, I was engrossed by chlorophyll and the miracle of photosynthesis, which allowed plants to grow from sunlight. My mind seemed to be in complete harmony with the breathtaking nature that enveloped me. Extraordinary emotions, which I inscribed in my journal, were authentically sealed with tears on the lined pages.

The vigour gained from feeding on protein incited me to set me off exploring much further than before, beyond known territory. I crossed the hill to a small lake and beyond to the next, and climbed the highest mountain in the area, gaining a magnificent panorama. No fruit though, so I proceeded down the opposite flank of the mountain with the intent of making my way back to the left. I continued along the berry path, which led to a dark valley, at the end of which stretched out an inviting lake.

Finding nothing noteworthy, I swung to the south, heading toward the sun that occasionally pierced the clouds, in order to return to the reservoir. After an hour and a bit of hiking, still no sign of it. Tired and somewhat irritated at having ventured so far, I took a break on an erratic boulder. Still facing south, I appreciated the solar heat tanning my face. Facing south? What was I doing walking south? The reservoir was to the north! Shit! How could I commit such a ridiculous mistake? Especially

The Philosopher

since I had lined up the polar star just about every night and knew perfectly well in which direction it lay!

Frustrated at this major waste of energy, I turned my back to the sun and headed north. But then the sky became overcast, and all the hills and mountains looked similar. Had I come this way? Nope. This way? Not sure. The sun briefly pierced the clouds and was now on my right. Oh my! I was going around in circles. I was totally lost! Yikes!

For an instant, the idea of spending the night alone and without fire set me into panic mode. But I resorted to my usual technique, marking the spot where I was at by erecting a huge tripod, and then searching in a structured manner, always returning to the temporary teepee, the only place in the vast wilderness from which I knew I could not be more than an hour and a half from camp. After exploring a half dozen futile directions, I came upon a weird-shaped rock that I knew I had seen before. Without ever losing track of this familiar landmark, I searched some more until I happened upon the sandy spot where I had captured the grouse. Known territory. Phew! I arrived at the reservoir just before dark and fumbled my way along the shore back to civilization, or so it seemed under the circumstances.

Another highlight of our last days in the forest occurred as we were peacefully lying in bed, almost asleep. Suddenly we heard a wild howl, followed by another, then a whole choir yelping in unison. Wolves! They were but a hundred metres from us, as the tracks found the next day revealed. I knew that wolves would not attack humans, especially sitting by a fire, yet I could not help shuddering. Awesome event!

My bow ended up being a fine piece of workmanship, if I do say so myself. The key-tipped arrow would easily stick into a stump at target distance. Too bad there were no animals to shoot at. Even Jacques's beaver had relocated to safer grounds, away from our hungry stares. I also built a Paiüte deadfall, a trap meant to drop a weight onto an animal, baited with the sucker's remains. Two days later the trap had not budged. I tossed a nickel on the sensitive trigger, satisfied when the logs came crashing down. I didn't bother to set the trap back up, but should have, because the next day the stinking fish was gone. I swore I would learn more about trapping, and did obtain my trapper's licence a few years later.

As far as mushrooms go, rock tripe did not grow in our territory. I found a great many specimens of the *boletus* and *suillus* genera, but practically none that weren't infested with maggots or slugs. We did enjoy many substantial skewers of large peppery mushrooms called *Lactarius deceptivus*, after hesitantly swallowing minute pieces for a few days. The problem was that I couldn't remember which of the edible *Lactarius deceptivus* and the poisonous *Lactarius vellareus* had forked gills, the only differentiating factor between these two closely related species.

On the third day before last, I was convinced that fruits were no longer sufficient to maintain my health. I was famished and gulped down berries with surprising greed. I felt I was just on the verge of this change in my psyche that would permit me to eat normally unacceptable food, such as insects. Even if I had long before unsuccessfully attempted to eat water-lily roots, their pretty yellow flowers enticed me to hold my breath and dive into the murky waters of the pond to excavate a few arm-sized pieces. But even after boiling cubes of them for hours in five changes of rusty water, they still tasted like concentrated pickle juice, and in spite of my ravenous appetite I couldn't swallow a single bite. Once again, I cursed those despicable armchair survival charlatans. As the ink of the night invaded us, I dragged myself to the lake. Using a birchbark torch, I illuminated the black and mysterious water. No fish, no tadpoles, no frogs, no crayfish. I was hopelessly out of ideas.

A lost tooth filling provided me with another occasion to philosophize. After I had wedged a piece of spruce gum in the hole, I remembered vividly how during my travels I had witnessed this poor African fellow who suffered from an abscessed tooth. Without access to a dentist, I watched helplessly for two weeks as his mouth and cheek swelled with pus until the tooth fell. I reflected that it wouldn't take much here to raise my standard of living above that of many Third World countries. A few tools, nets, fishing lines, soap, insect repellent, matches, tarp, and sleeping bag would do the trick. In the end, I couldn't even understand why this Survivathon should attract attention when much worse could be seen every day in disadvantaged countries. I concluded it must be because I was born with white skin, in a rich country.

The Philosopher

Mid morning on August 31st, a helicopter landed at our site and out popped the media to feed our ego. For several hours they bombarded us with questions and filmed our routine, asking us to pose as actors while gathering wood, poking the fire, lying in our beds, sharpening keys, fishing with the snare, hunting and berry picking. Hollywood! Flies had

Jacques and I posing for the camera on day thirty-one.

become so rare that I often went about shirtless while near the breezy shore, yet the camera crew complained of bites during their entire stay. Near suppertime, they helped us compensate for our lost time by joining the berry-picking routine. It was agreed that they would bring no food, for we didn't want to falsify the scientific data that was to be recorded as soon as we set foot in town. And then they were gone, leaving us alone for our last night.

An hour later, as I promenaded gently on the water's edge, pensive, a shiny reflection between two logs caught my attention. Huh? Empty cans. Garbage. In my paradise! No insult could have been worse at that precise moment. It was a knife straight through my heart. Barbarians! For an instant I considered never going back to so-called civilization.

A witchy storm was brewing. We sought refuge in our Noah's ark. Jacques and I high-fived, considering it would float us through the deluge. We had made it. The Survivathon was over.

But the gods decided otherwise. As if to show us who was boss, Thor himself threw his hammer at us that night. He proved our shelter was nothing but a sieve in disguise. We were drenched to the bone and didn't sleep a wink. The echoes of the tempest's mocking laugh made it painfully clear that although we had survived thirty-one days, Nature was still master in her own home. Even our fire barely resisted the flood, despite the fact that I had covered it the best I could with a flattened log. Never in my life had I felt so humble, so very small.

The new day left us anxiously wondering if our helicopter would fly. I felt quiet, serene. On one hand, I was eager to return and to hop into a luxurious bath. On the other, I felt concerned that I would lose the unique feeling of appreciation I had gained. From nature's perspective the trees don't grow in the city, it is the city that grows around the trees.

I also felt a sense of helplessness against the enormous task of reversing city ideologies. Blackflies provided the answer. A single fly is but a speck that can be crushed between two fingers. But a horde of them can reduce the toughest redneck to mush in mere minutes. Similarly, the hands of the powerful can crush a lone individual. But, as member of a larger lobbying group, no government, no tyrant can prevent justice.

The Philosopher

The runaway son was about to return home. I wondered how he would react to the first sign of civilization, no doubt the polluting chimneys of Saguenay's' major industries. Would I gather a minimum of equipment and return to nature, as so many hermits had done? Or would I learn to adapt, taking only the best of both worlds and helping the best I could?

A motor's noise interrupted my thoughts. The helicopter! We grabbed bark from our roof and boughs from our bed and tossed them on the fire as a signal. An impressive column of smoke rose to the sky, followed by a sparkling fire. Damn! The noise wasn't the helicopter, just a distant aircraft passing high above. A fine mess we had just gotten ourselves into, if the helicopter was to be delayed.

My watch read eleven thirty, precisely thirty-one days to the minute since our arrival.

I spent my last hour reflecting on my life, my aspirations. What was my purpose in this magnificent universe? Did I want to raise a child someday? What kind of world would we have to offer him or her? I decided to write my future offspring a letter, right there and then, as a gift from nature, to be offered at the age of reason.

Somewhere, far away in the forest, September 1, 1984.

Dear child,

We brought you into this world at a time when planet Earth is in crisis, I agree. Growing up, you will often feel mixed-up, frustrated, discouraged, even appalled. You will look for your personal reason to live in a sometimes incomprehensible, illogical, if not absurd world. All kinds of stimuli will attract and influence you. Be on your guard as you'll need all of your intelligence to avoid life's pitfalls.

I'd like to give you the most precious gift of the universe, the spirit of nature. Grand is its harmony. Walk its true path and you will find joie de vivre. Because nature, always wonderful — although sometimes tender and sometimes severe, will always answer your questions, will make you laugh and perhaps cry, but will never disappoint you.

In concrete terms, what does it mean to live in harmony with nature? Well! Here are some tips:

** Enjoy the daily miracles of life. Slow down, observe birds, plants. Take time to taste, to distinguish the subtle flavours of food. Take pleasure in your shower, soap, your bed. Appreciate your home, keep it clean and attractive.
** Enjoy intimate moments with your friends. Do something special for them.
** Money will never bring you security. True wealth is knowledge. Money will never bring you happiness either. Real joy resides in love and tenderness.
** Possessions are but a gold-plated chain. Simplify your life by limiting yourself to real needs.
** Smile at people around you, regardless of their race or colour. Travel to impregnate yourself with other cultures. Be positive. Have fun. Forgive easily.
** Care for your health. Eat simple, healthy foods; avoid chemical additives. Keep yourself fit with daily exercise. Whatever you do, don't let anyone encourage you to smoke; this is the worst with habits. And be reasonable with alcohol, so it does not take control of your life. And drugs? They will prevent you from embracing real life.
** Reach your decisions with fairness and integrity. Don't blindly follow others. Only you have the power to decide what's best for you. Always look at things and people as if you were seeing them for the first time. Use your creative power to continually reinvent.

The Philosopher

- Don't be afraid of failure. It's the only way to learn. Between the new, which is uncertain, and the old, which is stable but stagnant, always choose the new.
- Make sure that your work is a contribution to the well-being of our planet. Fulfill your potential; this is the key to happiness. Value non-remunerative work as much that which is paid; judge work in terms of its intrinsic rather than monetary value. You may wish to learn how to cook your bread, sew your clothes, grow your own vegetables.
- Seek silence.
- Participate in the conservation of energy movement: reduce, reuse, recycle. Prefer non-motorized transportation, even if it takes more time to get there.
- Struggle actively for peace and justice in the world. Join groups working at promoting these values.
- Protect the environment at all costs. Join people who work at saving wild places.
- Learn all you can about nature. Discover plants and animals. Learn how to dress properly and comfortably in the outdoors. Enjoy tiny flowers and insects, butterflies, rich colours, new odours, and new tastes. Watch how dew on grass shines like diamonds. The most striking human creations are nothing compared to the miraculous beauty of nature.
- Avoid the numbing effects of television, which will influence you way more than you might think. Instead, set an objective to learn something new each and every day. The joy of living is directly related to learning. At least, judiciously select the programs you watch.

My child, stay motivated, your contribution is important. I hope that your generation will be able to repair the errors of mine. There is still hope. Long live nature!

I kiss you tenderly.
Dad.

Just as I was putting down my pencil, I heard the helicopter, the real one this time. An emotional teardrop glanced off the paper. I joined Jacques at the water's edge. We embraced hard. I filled my hat with water and rushed to extinguish our precious fire. I strongly inhaled the emanating steam, savouring the moist heat on my hardened face. But the helicopter was waiting.

Twenty-One

Retro-Propulsion

"Study the past if you would define the future."
— Confucius

Jim the pilot was inventing eternity as he filled the gasoline tank with jerry cans that he had transported as baggage. The regional newspaper was lying on the back seat of the aircraft. We had made the headlines, complete with a full-page photo of us with arms victoriously outstretched. I devoured the words until prevented from doing so by the whirring of propellers. We asked Jim if he could circle above our camp one last time to impregnate ourselves with the territory in which we had survived for thirty-one days. What a sweet impression.

I immediately enquired of Jim whether he had any matches. He handed over a partly used booklet of paper saviours, which I cautiously stowed into my pocket. Now we could crash, we would survive. Stopping briefly at the heliport of the Chicoutimi hospital, we threw ourselves in the arms of our friend Jean-Claude. He had arranged for a doctor to briefly examine us prior to hopping over to the parking lot of the city mall. I had expected a hundred people; there were thousands, all waving at us.

When we landed in the crowd I was looking for a single pair of eyes, those of my mother. As soon as I located them, I pointed a finger and rushed toward her. For a brief moment, the multitude disappeared. After

I had also embraced my father and my brother Réal, the swarm was suddenly back in focus.

I had sworn to deal with each individual as a special person, and so with utmost sincerity I saluted each with care. Many close friends had come to greet me. I was moved. Women, especially, each more beautiful than the other, greatly impressed me with the bright colours of their now seemingly foreign costumes and the strange scent of their perfumes. One spectator surprised me with her remark:

"Mr. Bourbeau, what you two have done is so fantastic, so extraordinary, it's more important than the coming of the pope next week!"

Another individual was so insistent that I give him a piece of my shirttail that I offered him my sleeve toque as a gift. He was jubilant.

Jean-Claude was carving a path in front of us to inside the mall where a press conference was to be held. We eventually climbed the podium, and Jean-Claude lifted our arms as a referee does with winners in the ring. Speechless, Jacques and I bear-hugged. Adrenaline-driven and without thinking, I grabbed my buddy by the waist and lifted him straight up over my head.

The post-Survivathon press conference.

Chairs were provided for the two of us, as well as for the president of the foundation for Nature's University and the rector of the University of Quebec. The latter was elegantly dressed in full three-piece suit and tie. What a contrast to our rags!

After the dignitary speeches, our responses to the journalist's questions had the crowd roaring with laughter, like when they asked what we ate and I related the burbot incident with an exaggerated imitation of the sucker's ugly mouth. I appreciated that people were so attentive we could have heard a bug fly. Gradually the crowd dispersed and left us with our friends. I excused myself a moment to go to the bathroom. A mirror! What a shock to see my thinned face wearing those spectacles. Soap! I remember watching the muddy water spiralling down the drain.

It was nearly four o'clock before Jean-Claude drove us over to the university's physiology laboratory. My colleagues measured our anthropometric data, took blood samples, and evaluated our fitness levels. Not having eaten a morsel since the previous day, I couldn't believe I was still able to execute forty-eight sit-ups in a minute. The data revealed that I had lost 9.4 kg and Jacques 13.6 kg. Only our electrolytes were below norm from the lack of salt. Aerobic capacity had taken a drop due to our intense fatigue. After a few days both measures had returned to normal.

Finally we were each offered a hotel room where we enjoyed the long-awaited shower. What a thrill. Then I absolutely devoured four bowls of mom's homemade chicken soup, with dried fruits, nuts, and a sandwich. All this I considered appetizers before our subsequent smorgasbord at the restaurant. I was as amazed as everyone else by my appetite. I started with another soup, a veal steak with rice, and two orders of fondue. Then I ordered a fried chicken breast with a portion of mushrooms in garlic butter. After two raids at the salad bar I was finally ready for dessert. I polished off a slice of apple pie, a butter tart, as well as a piece of pecan pie. I was stuffed, but the management brought out a specially prepared chocolate cake, on which they had erected a primitive camp made of coloured fondant. I honoured the chef by eating two pieces, which was hardly a chore. Mom couldn't believe her eyes. She never found out that I had gotten up in the middle of the night and drove to town to fetch myself a giant poutine.

The next day I woke early and headed to the restaurant again. I couldn't wait to eat! Eggs, bacon, sausages, toast, pancakes, jam — I ate everything in sight, even the remains in the plates of other customers sitting near me. I could not tolerate the waste of food.

After three days of such a regime, I had gained back 6.8 of the 9.4 kilos lost during the month! And I was never ill; perhaps the giant fish and more abundant berries of the last week had prepared my stomach for the feast.

This phenomenon of extravagance also occurred at the intellectual level. Thirsty for action, I plunged into life with a passion. That autumn I presented the Survivathon research results in scientific circles, gave in excess of thirty public lectures throughout Quebec and Ontario, appeared in over twenty television programs, produced a slide show, worked on a video, began to write a book, revised my computer software for trip food, and developed a prototype survival kit for the Boy Scouts, all the while teaching three courses at the university. I couldn't understand how students could complain about the small amount of homework I asked of them.

It took until spring before I calmed down.

Meanwhile, behind my back, Jean-Claude had submitted the Survivathon adventure to the review committee of the Guinness Book of World Records, who accepted to log the trip as the longest ever voluntary wilderness survival experiment. That, plus the book I wrote in French relating the day-by-day details of the *Surviethon* turned me into a public figure in Quebec, whether I liked it or not. At first I enjoyed the spotlight, but I soon tired of it. After a hundred re-iterations, I started cringing when approached or introduced as "The guy that spent a month in the bush. You know, the guy with the funny wooden glasses." Or even worse, "You know, the guy that ate squirrels." Apart from those reminders, though, the Survivathon was behind me. I was relieved when the years erased memory of me so that I could walk incognito again.

Thirsty for a new project and using "Semester at Sea" as an example, I convinced the university's administration to let me offer its first ever multiple-course expedition. It seemed to me that the curricula of outdoor leadership, research seminar, pedagogy, and philosophy of

Retro-Propulsion

outdoor education, and a special project in physical education could all be combined with advantage. A group of dedicated students accepted the challenge. We dubbed the alternative education endeavour "Changing Gears '86," an appropriate name for a cycling trip across Colorado's Rocky Mountains and over to California and the Pacific Ocean. After a year of preparations and fundraising, we clambered aboard the mini-bus on our way west, bikes and panniers in the towed trailer, and drove across eastern America.

Part of the course work included planned visits, and we stopped to tour the manufacturers of outdoor equipment such as North Face and Marmot Mountain Works, received lectures from the leaders of the outdoor programs in multiple universities, participated in summer camp programs, toured a nuclear power plant, and so on. The research seminar tested bike equipment and food packing systems. We carried a rolling library of books on outdoor philosophy, with daily readings. By necessity, one student learned to cut hair. Intensive English immersion was a bonus. Outdoor education pioneer L.B. Sharp would have been proud.

The trip rolled along on ball bearings, both in the real and figurative sense. Until one night. Sleeping like a log in my tent after a hundred-plus kilometre day, ear muffs on, in a standard private campground, I was startled awake by the blaring sound of loudspeakers right next to me and an ultra-bright spotlight shining on my tent.

"Are you the leader of this group?" yelled someone through the microphone. I stuck my head through the tent door and stared down a revolver barrel aimed straight at my squinting face.

"Yes I am, what in the world is going on?" I nervously answered with my hands up, kneeling in my underwear.

And the uniformed cop, backed up by three others replied: "Your boys were yelling murder threats to the camping neighbours!"

Turns out my team had been partying with alcohol and I hadn't heard the slightest sound. And now of course, they were all dead asleep and I couldn't raise a soul to answer my questions, nor the cop's. Since they were partying in French, I suppose the worried neighbours had imagined the worst. After plenty of apologetic explanations, I got off with a warning.

The next morning I faced a very timid group of students who stammered regrets at having bypassed the rules they themselves had written up and formally signed, stating that they would restrain alcohol consumption to below legal driving limits. I had learned a good lesson as outdoor leader: you can't override human nature. From then on, all long trips near stores would include supervised party nights with paid security guards.

We had many minor adventures on the trip, like a flipped raft in rapids and a scratched helmet following a gear-failure-caused accident. But the highlight was certainly climbing for hours up and over snowy 3,448 metre Monarch's Pass, then sliding down the other side through mushy sleet.

Project "Changing Gears '86."

Once back in Chicoutimi, and the post trip tra-la-la taken care of, I resumed my work being a typical university professor for a while, which was comprised of three main tasks: teaching, research, and community services. So I wrote and compiled course notes, organized a scientific colloquium, submitted memoirs to protect wild rivers to the Bureau d'Audiences Publiques sur l'Environnement, contributed to Expo-Nature,

Retro-Propulsion

wrote a chapter on survival for a collective book, prepared and submitted a few articles, and so on. Tons of boring stuff.

Then one day I received news through internal university mail that grants were being offered to researchers for projects related to the 150th anniversary of the Saguenay-Lac-St-Jean Region, two years hence. What better way to celebrate such an event than to relive 1838? Why not show the population of the Domaine-du-Roy what living and travelling was like when twenty-one individuals founded La Société des Vingt-et-un and came to colonize the fjord?

After some lobbying, I obtained the funds to organize a historical re-enactment expedition that would cross the entire region, stopping in every town and village (over thirty) to perform demonstrations of our ancestor's way of life. After spending days at the National Archives, I unearthed the original journal of Joseph-Louis Normandin, the first surveyor of the region in 1732. Normandin had travelled from the Metabetchouan fur trade post on Lac-St-Jean, up the tumultuous Ashuapmushuan and Chigoubiche Rivers, along the Licorne River, across a chain of lakes, and then up the river that now bears his name to end in the Nicabau fur-trade post. I had found my itinerary. As authentic "coureurs des bois," we would follow Normandin's route to Nicabau, pick up some bales of fur there, and lug them back along the same route and onward downstream on the Belle and Aulnaie Rivers, across the chain of lakes to the Chicoutimi River and down the Saguenay River to Tadoussac on the shores of the majestic St. Lawrence River. Nice.

Today, of course, organizing such an expedition would be simple enough. Since the advent of Internet, a great number of online stores have popped up where one can buy accurate reproductions of each and every piece of gear needed. But back in 1988, we had no choice but to make our own. Just trying to figure out what gear was available in that time period and what it looked like was a major undertaking. And since this was a research project, I wanted everything to be scientifically accurate. No compromises. I had no idea what I was getting myself into.

"How much for those cow guts?"

"You can take as many as you want. No charge."

"Wow, thanks!"

My long time buddy James Déraps and I grabbed a dozen soggy bladders and a few dirty intestines and exited the murderous scene of the slaughterhouse. Back at the indoor locale of the university's "outdoor lab," Mike Martineau and Marcel Savoie, two former students turned good friends, were busy hand sewing their smoked rawhide moccasins with a homemade awl. James and I washed out the slimy guts and hung them to dry. We figured these would come in handy to keep our stuff dry.

"*Merde*, those things stink!"

Mike was right, so we opened the window and hung them outside, dangling from a rope draped over the sill and tied to a central heating radiator. I wondered what the passers-by would think. Who cares! I finished cleaning the sink and threw the remaining gut scraps into the aquarium, where Tourtière, my loyal snapping turtle, resided.

I glanced around the room, satisfied with the results of our last nine months of work. We'd forged our crooked knives and needles from old files, retraced Normandin's original map on canvas with India ink, sculpted a checkerboard onto one of the many paddles we'd crafted, found a copper pot and dishes, and bought a Brown Bess flintlock musket. Since I couldn't buy spare flints for the gun, I went to meet a flint-knapping pro in the States who showed me how to make them. In fact, most of the information we'd gathered had been through word of mouth contacts, and I'd been enjoying getting away from Chicoutimi to visit the people I was referred to.

We'd also hand sewn countless ditty bags, gathered spruce gum and extra bark as a canoe repair kit, waterproofed cotton tarps by painting them with raw linseed oil mixed with turpentine and iron oxide, found an elderly lady who showed us how to make bars of soap made from fat and ashes, bought some intricately decorated birchbark vessels from a Native artist, and scrounged up pure wool blankets from the back shelves of various flea markets. On the central table were traditional wool socks, containers of natural bug dope concocted from camphor, citronella, and cedar oils, a period comb, four tumplines, two types of gun powder, bear grease, a fishing net, noggins, a brass compass, coils of rope, a leather-bound journal, ink, vermillion powder, an axe, and a crude pig-bristle toothbrush of my own undertaking. Oh, and a small wooden barrel to carry the rum, of course.

Thanks to my research assistant and friend Trudy Morphet, also a wonderfully talented artist and craftsperson, the hand sewing of our undergarments, felt pants, and cotton shirts was coming along fine. I also felt deeply indebted to her husband Greg Cowan, who crossed Canada as a voyageur back in 1967 and who was helping us build gear. Greg happens to be the most incredible handyman I know, one of his accomplishments being the construction, with ancestral tools, of a fully outfitted historic wooden sailboat, complete with steam engine. I was hoping he would accept sailing our fur bales and us the hundred and twenty kilometres from Chicoutimi to Tadoussac at the end of our trip. He was visiting and in no time had helped me transform two raw bullhorns into fine-looking and functional gunpowder horns.

Most importantly, I'd convinced my friend and mentor Kirk Wipper of the Kanawa Canoe Museum to lend me two of his precious birchbark canoes. Although he assured me that those two particular canoes did not have historical significance, I was well aware that he would not let anyone else use them for a trip. He even told me the canoes would be worth more after they've been on a trip with me! Nice guy.

But there was still a lot to be done. My traditional fletched sash belt was in the process of being woven by hand, knot by knot, by Mr. Rondeau of the Association des artisans de ceintures fléchées. He was charging me eight hundred dollars for an eight hundred hour job. A passionate artisan! I still hadn't solved my eyeglasses problem and we hadn't even started packing food.

I'd been spending a lot of time with the region's historians gaining insight on the way of life of the coureurs des bois. But to understand their concrete day-to-day life, it was my Native mentors Jimmy Bussom and Gérard Siméon I turned to. Their knowledge base astounded me; at my first contact with them, I became aware of my beginner status. Humbly I soaked up their teachings. Through them I met René Robertson, the Native fur dealer, who set me up with a large batch of sub-category fox, wolf, and beaver skins. The historic fur-trading post and tourist attraction in Metabetchouan would furnish me some fake bales of pelts for the return trip.

After having spent over a year of spare moments reading every historic journal I could lay my eyes on, I surprised myself by feeling transformed, perhaps like an Elvis impersonator who starts believing he actually is the King. As I read Robert Ballantyne's account of his service with the Hudson's Bay Company in the 1840s, his description of himself in the mirror after a trip served as model for my own clothes: a blue canvas capote, brown corduroy pants, a headband, a sash belt, and Indian moccasins. To this outfit I had to add eyeglasses. In those days, you looked through a series of demonstration lenses at the trading post to pick out the ones that corrected your eyesight best, ordered them from England, and received them a year later. The Saguenay Museum happened to own a pair of original frames from that time period. I copied them exactly by welding pieces of metal tubing cut lengthwise into circles. The side bands didn't reach behind the ears, they ended in rings near the temples instead. To those I tied a lace of eel skin, just like in the old days.

Three months before the trip I hired two bright gals as a promotion team, Jeannette and Jacynthe. They planned all of our demonstration stops and handled the logistics and the media while the expedition was in progress. During a brainstorm, we decided to call the project Retro-Propulsion, the French term for back paddling. Since the stroke was so useful to descend rapids safely, and has the back/forward double meaning, we found it was particularly appropriate for a sojourn whose aim it was to paddle in the past to better foresee the future. A way of introducing the save-the-rivers message.

Since all was going according to plan, I also wrote up a full description of the project and sent it off to National Geographic and a dozen other similar organizations in the hope that one of them would be interested in filming the story.

One month before departure I contacted one last time the Masteuiash Reserve to try to convince the Native people to join me for at least part of the trip. In a fit of extravagance, I suggested they could travel with us on Lac-St-Jean in a *canot du nord*. They said it was possible. Getting Native

Retro-Propulsion

The resuscitated Robert Ballantyne.

people involved seemed like a wonderful opportunity to share with them, so I called up my friend Kirk at the canoe museum and begged him to lend me a large birchbark vessel, which I would take care of like my own eyeballs. He agreed.

The next weekend I hopped in my Volvo with James and we drove the thousand-kilometre trip to Ontario to fetch the canoe. Professor Kirk could not meet us at the museum, so he told me where the boat was and authorized the security guard to give me access. When we arrived, James and I were stunned at the size of the beautiful craft. There was no way to load the thirty-foot canoe onto the car, unless … We headed off to the lumberyard and bought a dozen two-by-fours and a hammer with some nails to build a huge rack. An eternity later we were finally off to Chicoutimi, but couldn't drive more than seventy kilometres per hour. For eighteen long hours, every single vehicle on our route stared at us in disbelief. We decided to deliver our unusual car-top load directly to the shores of Lac St-Jean, an extra three-hour drive.

We arrived home by sunrise on Monday morning, and I gleefully hopped into bed to catch some sleep before my afternoon class. But the phone shook me awake at 10:15. It was Kirk. We had grabbed the wrong canoe! The one we had wouldn't float; it was built for museum display only. Oh well. It would make for good photo shoots, I guess.

The final preparations went well as we packed up our gear and food such as peas, barley, homemade hardtack biscuits, whole-wheat flour, and dried prunes. Oh yes, and one whole smoked beaver that my Native friends had graciously provided, warning us to keep smoking it as we went.

Then Jeannette popped in with great news. KEG Productions/Ellis Enterprises from Ontario had agreed to send a film crew to spend ten days shadowing us in the wildest part of our journey. And not just any film crew. Ken Buck would be behind the camera, the very guy who filmed the legendary Bill Mason in his red canoe. Wow!

I phoned the director Ralph Ellis and we agreed that to maintain scientific integrity we would have no interaction whatsoever with the film crew, who would discreetly follow us from a distance.

At long last it was D-Day. The four of us proudly posed besides the huge north canoe, which would only serve on our return leg during the

Retro-Propulsion

An unusual car-top load.

The Retro-Propulsion crew posing, just prior to departure.

promotion/demonstration segment of the trip. Then we waded into the water and loaded up our fragile flotilla. Mike was in the bow of the fourteen-foot canoe I was sterning, whereas James sat in front of Marcel in the sixteen footer.

The inland sea is pretty rough for such small and overloaded craft, and we barely progressed one kilometre before we had to pause to reconsider. Our canoe was shipping water over the side, and Marcel's and James's canoe had already sprung a leak. We stopped for repairs on a private beach and had to convince the owner we were not weirdoes. Near suppertime the winds abated and we pushed onward. It was a particularly scorching July day, and soon thirst drove us to pull over again, for the lake was too polluted to drink. We used the flintlock musket frizzen and gunpowder to get a lamp wick smouldering, and then blew the cotton into flame using shredded cedar bark kept dry in a cow bladder. We set a pot of water to boil and as soon as it cooled to lukewarm, gulped the contents.

For three days we had to repeat this process bi-hourly, as we paddled past the towns of Roberval and St-Félicien and started up the Ashuapmushuan River. What a relief it was to leave the Donahue pulp mill behind us, and finally attain clean water.

Clean water, but disgusting food. At lunch we discovered that the body cavity of our smoked beaver was being devoured by hundreds of maggots. To think that last night, in the darkness, we were feeding off of this same carcass. Oops! We should have listened to our Montagnais friends.

One of the canoes was giving us endless trouble. In desperation I removed half the ribs and cedar planks and slapped pitch on it from the inside, covering the leaky spot with a tarred rag.

We were following Normandin's daily journal entries, and we were already a couple of days behind his 1732 schedule. Portaging became unbearable because we were carrying too much junk. So when Jeannette and Jacynthe came to meet us at Chutes à l'Ours to check on our progress for a media report we unloaded just about everything except food, one pot and noggins, our tarp, one blanket, and an extra shirt each. We entered the wild Ashuapmushuan Reserve, and wouldn't have contact until we reached Chaudière Falls, two days travel according to

Normandin's notes. Ken Buck and his canoeing partner joined us and were following behind, tiny dots in a Grumman aluminum canoe. We soon forgot they were there.

The mosquitoes must have known exactly where the boundary of the reserve lay, for as soon as we entered our first wilderness portage they proved without a shadow of a doubt that our ancestral bug dope was useless. That night I draped some cheesecloth over four sticks poked into the ground to create a bug net over my face. But in the morning as I woke to a smiling camera peering down at me, I noticed that the enclosure was jam-packed with the needle-nosed critters. In the wilderness, it was easier to make friends with our brother the wolf than with our cousin the skeeter. These distant relatives were having a happy party this morning at my expense, with shelter, warmth, and an all-you-can-eat skin buffet.

Ken was discreet. No sooner had he filmed his shot, he was gone. We noticed him again later as he crept up to capture images of us waiting out a downpour under our tarp. He didn't say a word.

During the summer, the Ashuapmushuan flows at about three hundred cubic metres per second, and the current rushing against us slowed us to a crawl. For four full days we paddled hard from dawn until dusk, unless we were portaging, lining, or dragging the canoes, sometimes with water up to our waists. Late afternoon of the fifth day we reached the base of the Epinette Blanche Rapids, a kilometre-long R-3 without a portage trail. We attempted to go up the right bank. Wrong choice. After a couple of hours of dragging the canoes against the forceful flood, we came to a tumultuous rolling wave at the base of a sheer cliff, which decisively barred our way. We would have to return downstream and ferry over to the other bank to try again.

But the day was way late and we were way beat. Trying to ride the waves downstream seemed suicidal. Out of options, a barely climbable narrow ravine was the only way to leave the river's edge. In semi-darkness, the weary voyageurs used ropes to lug their gear and their canoes up the crack and tie them to trees. We each found a spot to sleep by wedging ourselves between a tree and the rock face. Never again would I complain of a sloping campsite! And in comparison to our cold and unbreakable biscuits, any warm supper would be welcome.

To make matters worse, poor Mike was suffering from a serious case of poison ivy, which had spread from his legs to his upper body. Unfortunately there was no way to relieve him until we finally arrived at Chaudière Falls two days later. Eight days! It took us eight full days to cover what Normandin had managed in two. We bowed our heads in respect.

The daylong break at the falls was welcome, as we admired the giant rock cauldrons carved out over centuries by water-spun pebbles. We also appreciated the moose steaks our land team surprised us with. It sure beat the sea gulls we added to our barley the day before.

The next leg of the journey involved ascending the Chigoubiche River, starting with a steep portage over a gushing tall waterfall. Lovely. Judging by the overgrown and dense forest on both sides, it seemed as if the route had not been used since the fur-trade era. We grabbed the axe and took turns chopping a trail until we reached calm water. Since the Chigoubiche is profiled like a staircase, it was easier going than the Ashuap. Soon we reached Chigoubiche Lake and portaged two kilometres through bush over to the Licorne River, which was nothing but a meandering creek. When we reached Lake Ashuapmushuan, we attempted fishing with our net across its mouth, without success. Since this practice was illegal without the special permit we had obtained, we of course had no idea how to go about this. And with not enough time to learn by trial and error. Another question for my Native mentors.

Footwear was becoming problematic. We were not even halfway through the trip and our moccasins were pierced right through — not on the heels and toes, as expected, but rather in the centre of the soles because of the round cobbles we kept hopping on. The leather stretched way out of shape while wading, and at night the mocs shrivelled up to the point where we could hardly get them on in the morning.

One of my favourite pieces of gear was the greased-leather dry bag I sewed from a dog skin the local taxidermist had offered me. Even when submerged, my blanket and shirt stayed bone dry. I loved my sash belt too. It acted like a weightlifter's belt when portaging, made a suitable emergency tumpline, and doubled as a comfortable pillow. More importantly, because it was wide and made of wool, it acted as a warm layer when draped behind the neck, criss-crossed over the torso, and tied around the waist.

Retro-Propulsion

That night we were camping on a large rock outcropping. I ogled the oily water, wishing I could jump in to evade the heat wave. But in the calm before the storm, mosquitoes were particularly fierce; in fact, they were more ferocious than a junkyard dog. Tired out by the portage bushwhack, we decided to skip cooking supper. Starved, I dumped out a lumpy ditty bag full of walnuts onto the rock. Oops, that was the lumpy ditty bag full of spruce gum. I half-heartedly fixed my frustrating mistake, repacking just the biggest hunks of resin.

We decided to build a triangle of three smoky fires to get rid of the beasties, but they zoomed in on us anyway. It was hell. We lay down to avoid suffocating, with our blankets over our heads to prevent bites. The heat was unbearable. Sleep was hard to come by, but accumulated fatigue eventually forced my eyes shut.

I must have been dreaming. As if tarred and feathered, I was hot, sweaty, and gooey. I was rolling around in warm viscous glue, my hair welded into lumps. I pinched myself. I was awake. This was real! What the? Jesus! The heat from the fires had melted the spruce gum I had left laying on the rock, and it had flowed down onto me. The icky paste was plastered all over my hands, face, hair, clothing, and blanket. Yech!

Three days later we had reached the height of land and our destination at Lake Nicabau where our loads of fur bales were waiting. At long last we started back and could now enjoy flowing with the current. The Normandin River rapids gave us a thrilling ride, a wonderful reward for the upstream efforts. Then we entered the headwaters of the Ashuapmushuan and were whisked along. However, since we didn't have life jackets we portaged many rapids that we would normally have shot through.

During a long straight stretch of this fun part of our journey, I unlocked that spiritual mindset whose key is repetitive paddling. I couldn't help but be amazed at how I was floating on nature, both literally and figuratively. Below my knees were cedar boards and birchbark that simultaneously grow and breathe around me. The roots and gum holding the canoe together exist right there along the shore. My linen sleeves had been woven from plants I have seen in a garden, and I have witnessed the entire transformation process. Same with my moose-hide moccasins. What an awe-inspiring moment!

And as I delightfully watched the other canoe drawing pretty ripples on the dark blue sheet, I was transported into an era of simplicity and direct contact with our priceless environment. Clean, fresh air, crystal-clear pure water, good friends. I marvelled.

As we zipped along, I came to view the wild Ashuapmushuan's rapids as diamonds on Princess Nature's tiara, with the Chaudière Falls as its crowning jewel. Each tree applauded her passage, and we, lucky part of the parade, were enamoured with her beauty.

In that context, who cared if our food stores were down to flour, onions, and lard? Content with our onion rings after twenty-eight days of mostly bliss, we were back at Lac St-Jean. We all admitted to knee weakness when we first encountered the bikini-clad demoiselles who had come to meet us at our first historic talk on the beach. So much for historical accuracy! Musket shooting demonstrations and fire-by-friction or with flint-and-steel felt weird amidst the audience of beer-guzzling and Speedo-wearing city dwellers.

We traversed the big lake in paddling sprees from demo stop to demo stop. Only the passage straight across Chambord Bay made us wish we

James and I during a retro demonstration.

had gone around. The canoe was acting up again, and leaking as fast as James could bail. Close call.

The only other noteworthy event on the way to Chicoutimi was paddling the historic Aulnaie River through the village of Hébertville, as it was quite obvious where its sewage ended up. That sure flamed our talk of the day.

When we arrived in the city of Chicoutimi a few days later, a strange feeling overwhelmed us as we portaged around the dam on asphalt, dressed retro as we were. Especially when we had to wait, canoe on shoulders, for the traffic lights to turn green. Anchored next to the downtown Chicoutimi bridge, Greg awaited us on his charming sailboat *Jamestown*. He beckoned us to hurry; tide was going out. We tossed our gear and the fur bales on board, and sailed off toward the east, boat tilted to the max.

The Saguenay Fjord impressed with its majestic cliffs and deep, dark waters. We spent three days in another century's heaven, cruising with full sails on a broad reach all the way to Tadoussac. Perhaps there was a lesson here. Hmm, wilderness without hardships. Interesting!

The sailboat Jamestown.

Twenty-Two

The Birchbark Pillow

"The truth is that our finest moments are most likely to occur when we are feeling deeply uncomfortable. For it is only in such moments, propelled by our discomfort, that we are likely to step out of our ruts and start searching for different ways or truer answers."
— Morgan Scott Peck

The Retro-Propulsion experience was but an introduction to the world of historical research, having left me with more questions than answers. Sure, I had proven or refuted some of my hypotheses. For example, how were the rope painters tied to a fragile birchbark canoe? I guessed that they would have to be knotted around a bundle of gear shoved in the bow and stern. But when we started our adventure, I had simply tied them temporarily to the tiny wooden crosspieces found at each end of the canoe. As the trip progressed, we pulled more heavily on the lining ropes. At one point I had actually swamped my fully loaded canoe and had wrapped the painter around an alder tree in desperation; the little crosspiece held up just fine. Later, I used a dynamometer to test a five-millimetre spruce root for breaking strength: forty kilos. There were sixteen wraps on the lashings of the crosspiece. It could withstand a pull of over six hundred kilos.

Other answers were harder to find. I kept wondering why Normandin had been so much faster than us going upstream, whereas we could keep up to his pace going downstream. Further reading suggested that maybe poling technique was the answer. My Native mentor just smiled: "Anyone looks good paddling down river, but it's against the current that you can distinguish the real canoeist." Gérard was too frail to teach me, but I found other old-timers in Maine that could. It was only many years later that I became friends with traditional Maine guides Alexandra and Garrett Conover who poled extensively. They in turn introduced me to Harry Rock, the world sport-canoe poling champion, and then I understood fully. Their inspiration sent me practising until I too could ascend class 2 rapids with ease while standing in the canoe. Mystery solved.

During the year following the Retro-Propulsion expedition, I had the opportunity to present eight conferences related to my findings. I had also proudly testified to the value of saving wild rivers in over fifty television and radio interviews. My implication in the region was rewarded with a few distinctions, the two most important being the Saguenay Ambassador Medal and having the annual Expo-Nature Environmental Award titled after me. An unexpected boost for my career, and, of course, not hard on the ego either.

The summer following, Ken Buck and the KEG Productions/Ellis Enterprises team returned to Chicoutimi to obtain missing images for the film they wanted to produce. So James, Mike, Marcel, and I donned our historical garb once more and had fun recreating some of the events of the summer before, this time playing for the camera. Every scene was repeated twice, so original versions in both French and English could be realized. The hour-long made-for-television show became part of the award-winning *Profiles of Nature* series. "Perspectives d'un coureur de bois," and its English alias "Man of the Wilderness" were distributed internationally, and KEG did such fine work that the documentary earned a bronze medal at the Houston International Film Festival. As a bonus that summer, I added to my repertoire a whole array of new jokes gleaned from the producer Ralph Ellis, a wonderfully talented storyteller.

To this day, I remain fascinated with the material culture of our ancestors; it has become a great leisure activity. And in my mind, when

I read the journals of such legendary birchbark canoe explorers as John Franklin, Samuel Hearne, Alexander MacKenzie, or Simon Fraser, I feel a sense of brotherhood.

The Saguenay River had been particularly good to me, so the next May I proposed paddling its entire length in a Rabasca canoe to a small group of students as a leadership project. Cold water and spring ice with seven-metre-tide variations made it a perilous journey, so we wore wetsuits underneath our warm clothing. All alone on the water, it was a memorable trip. No incidents to report.

The next five years of my life were rather strange. Concerning outdoor life, a complete change in paradigm saw me experimenting with the very best ways of making myself ultra-comfortable in nature using modern gear and efficient techniques. My new challenges included setting up a five-metre group tarp in "PI time," three minutes dot one four seconds, in three different configurations. Or tossing the food pack up out of bear's reach in the same time-frame criteria. Or finding twenty new ways of cooking bannock on a fire.

Wilderness survival research also continued to interest me, but in a different way. I was less prone to accept suffering, yet I felt there were still many wilderness secrets to be revealed and I would occasionally spend a night or two out without equipment. But my main concern had shifted to finding ways to save wild spaces for future generations.

The oceanographic researcher Jacques-Yves Cousteau fascinated me. With the success of his Aqua-Lung invention, he had obtained enough money to bring to fruition incredible projects with repercussions all over the world. This inspired me greatly, and I secretly fantasized about making a wild amount of cash to fund my own environmental research projects, without having to go through the standard peer-review process to fight for over-solicited grant dollars. Like most overachievers in their late thirties, I was looking for a way to "make it big." At least I would give it a good try. If I hit the jackpot, I could have a whole slew of research assistants working on ideas with me.

My first attempt at fortune was to market my tripping food expertise. I had developed software to computerize distribution of trip meals to outdoor camps which I had labelled the GB Tripping System™. It was

Paddling the Saguenay River.

comprised of thirty full meals, all with recipes of my own creation. Thus far, GB Catering had distributed hundreds of thousands of these meals, so I knew they were appreciated. Plus we had the equipment to mix huge batches of my recipes, like cornmeal bannock for instance, whose base was eighty kilos of pastry flour, forty kilos of cornmeal, and twenty kilos each of milk powder, egg powder, and sugar.

Brainstorming found the name Explo-Rations™ and I convinced my dad to let GB Catering produce four of my meals and promote their commercial distribution throughout Canadian and U.S. outdoor stores. I had sent four samples to *Explore* magazine, who were writing up a taste comparison article of all dehydrated foods on the market, and my beef stroganoff won first prize. Plus the three other entries came in third, fourth, and seventh. I was hopeful.

Taste the adventure!

Unfortunately, we did not have the marketing budget to compete with such giants as Mountain House Foods, and a couple of years and a hundred thousand sales later, the project was abandoned for lack of profitability.

Other gambles at finding oodles of money were related to the computer software business. Ever since microcomputers had been invented, starting with a Pet Commodore 8K with cassette recorder, I had jumped into the revolution with both feet. In fact, all those years I led a parallel life as a computer programmer, the reason being GB Catering. At first, encouraged by my success with the food tripping software, I had helped our family-owned business with simple data bases, such as client lists and supplier lists. But as the company grew, Dad asked me to program software first for accounting purposes and payroll, then to manage the entire warehouse stock. The first micro-computers were not designed to run such memory-intensive applications, and I truly struggled. I have vivid recollections of spending entire nights programming and debugging software, and even once sleeping on a sack of flour in the warehouse while data was being updated. Somehow I surmounted this weird kind of survival ordeal, thanks to the help of two other programmers I had teamed up with. And as GB Catering added its GB Food Supply division and grew to over five hundred employees with a fleet of trucks, a bakery, and a butcher shop, my interconnected Macintoshes kept a dozen secretaries busy.

I was programming using a computer language called Omnis, and had become an Omnis registered developer as well as a certified Apple developer. With my friend James, who was also a computer addict, we started up Info-Mac Enr. as a sideline, and our combined expertise served many of the region's companies. Through the grapevine, the owner of Softcode Inc., then the largest distributor of Apple software in Canada, heard about my multi-user Warehouse Control™ software. He was interested. Here was my chance to join the big leagues!

I spent that entire summer vacation and beyond sitting fourteen-hour days at a desk in Softcode's warehouse in Toronto, modifying my software for their own use. If they began using my Warehouse Control themselves, I couldn't even fathom how many they would sell. This was huge. By the end of summer we were ready. I patiently trained all their staff who were excited at the manual work the program would save them and over a weekend transferred all their existing data to the new software's format.

A small bug. A tiny small bug, somewhere in my software, was subtracting instead of adding tax. The boss stopped delivery for two hours, as I frantically searched for the bug. Then he pulled the plug. While they re-installed their old software from their back-up copy, I found the misplaced negative sign among the thousands of lines of code. But it was too late; their faith was gone. All I had earned for my troubles were a few fancy lobster meals at the Chinese Palace paid for by the boss.

In a way I was somewhat relieved that this cement-centered activity was over. I was beginning to feel obsessed about borrowing so much time from my university job. Yet, a few months later I was programming again, but this time to provide researchers with software to manage

The Birchbark Pillow

bibliographical references. I figured that since I needed this kind of software for my own research, other professors and professional investigators would too. Half a year later, with the help of two assistants, Références Contrôle™ was born, complete with manual. It was going to be a hit. Or would have been, had I known anything about marketing. Passionate as I was, I devoted all my time to adding new features to the software. Great software, no marketing. And then Apple decided to enlarge the Macintosh screens. It would have taken another major effort to upgrade the software. I dropped seventy thousand dollars into that venture. Sold six copies at $79.95 each. At least I was cured. Definitely, I would never be a businessman. That strange tangent of my life was over.

It just felt so right to be breathing fresh air again. As I sat in nature, free from business considerations, I could no longer comprehend what had drawn me to the world of asphalt. Oh yes. Finding money to protect

Software for monks.

our wild places. I had been like a mother reading about caring for her children instead of actually doing so. When you're up to your ass in alligators, it's difficult to remember the initial objective was to drain the swamp. I was sorry about that.

The only redeeming factor was that my work often obliged me to return to the wilderness to teach. It was always a pleasure to reunite with my first love, be it briefly. But I started frequenting her seriously again, resuming my search of her soul for secrets. And perhaps because I didn't tolerate the suffering anymore — at least not for long — I found many by looking in corners of her home I had previously overlooked.

These new overnight-only experiments didn't qualify as survival to me because I wasn't pushing my limits. Instead I preferred to think about them as simple camping-without-equipment outings. Often they were sparked by something I had read.

On one occasion I wanted to verify the value of the popular survival technique that suggests installing a wall of logs or rocks behind the fire to reflect the heat toward yourself. Searching the literature, I found the origin of this technique in *Camping and Woodcraft*, a book by author G.W. Sears, alias Nessmuk, written in 1880. On the first day of a camping trip, having forgotten his axe, he boasted of his idea to compensate by using his small hatchet to cut a poplar into four-foot lengths and piling them into a sloping wall behind the fire to reflect the heat to his bed. At that time, it was customary to sleep on a bed of fir branches, wrapped in woollen blankets, feet near a fire burning all through the night.

Since Nessmuk slept very well that night (without an axe!) he boldly stated that he had discovered the best way to keep warm. What he didn't mention was that it was perhaps a quiet and warm night. However, he did admit that long poles of firewood were abundant. Also, since it was the first night of his trip, his clothing and blankets were dry, which added to his comfort. In other words, if he slept well that night, it might not at all be because of the reflector wall. Yet, many early nineteenth century authors cited Nessmuk, and eventually the technique appeared as fact in later publications, even in the army survival manuals!

So one autumn night I amused myself by sleeping by the fire in my own back yard, thermometer in hand, measuring the temperature where

The Birchbark Pillow

I lay. Then I leaned a full sheet of plywood upright against two lawn chairs as a reflector wall behind the fire. The result? To my surprise, I obtained no difference, not even a single degree!

I repeated the experiment three times, always with the same outcome. Later, mesmerized by watching fat snowflakes float gently from the sky, a slight breeze chilled my backside. I instinctively switched the plywood and chairs to behind me to cut the wind. This was much better. I glanced at the thermometer. An increase of twenty-five degrees! Several tests later, I came to the conclusion that a wall behind was without a shadow of a doubt a major improvement to heat retention. From that moment on, I never wasted energy building a reflector wall on the other side of the fire. Rather, I erected a wall of rotting logs behind me, because they piled up easily, moulding to each other. I named this the "rotten-wall" technique.

For a long time I had been looking for solutions to alleviate the wet discomfort of the traditional bed of fir boughs, for on single-night trips there wasn't time to dry grass in the sun as I had done during the Survivathon. Setting up camp during a somewhat rainy evening, I had placed my huge pile of long firewood poles on the other side of the fire to dry. As usual, I was asleep on a bed of wet branches covered with birchbark, for lack of a better idea. I woke up soon thereafter shivering, which was brutal as always. After rekindling the fire I sat down on the stack of poles, taking advantage of their accumulated heat. Too tired and half asleep, I slouched uncomfortably on the pile of firewood for a few moments in a fetal position, delaying my return to bed. What a surprise it was to wake up three hours later! Racked and bruised, yes, but warm and dry!

On the next such outing, I decided to stack half of my firewood as a bed in front of the rotten-wall, placing the straightest pieces on top as a bed. Better to sleep on hard-and-warm-and-dry rather than on soft-and-wet-and-icy, I figured. I found myself relatively well until midnight, when my firewood ran out. The rest of the night spent burning my bed was no fun.

For the next few overnight experiments I simply gathered more wood to compensate. But one evening, it so happened that a rock lay at the foot of the tree on which I had leaned my rotten-wall. That gave me the idea, instead of using a good stack of additional wood as bed, to just take half a dozen of my straightest poles and lean their one end on the rock and support the other on a square cage built by criss-crossing short pieces. The result looked like a knee-high bench with a space below. What a discovery! I slept the best night of my survival-skills-seeking life that night. The heat of the fire radiated under the bench and up the rotten-wall behind me. It was like a reflector oven, with me as the roast. That's how the "park-bench" technique was born. Later I refined the execution by supporting the bench on two long poles sticking out of the rotten-wall above the second log.

Over time, and with fine-tuning, a pattern emerged for comfortable survival-style camping without equipment, which I repeated countless times and structured so as to better teach it to my students. I invented composite names for each technique to aid memorization of the step-by-step curriculum elements.

Rotten-Wall with Park-Bench in progress

The Birchbark Pillow

1. First, find a **flat-and-calm-spot** that is wind proof, using **x-ray-view** to find a dense area and a **count-to-ten-spin** to ensure sufficient firewood.
2. Then find two north-south trees on which to lean a chest-height **rotten-wall**.
3. Dig a **wind-trench**.
4. Respect the **no-touch-zone** while gathering materials.
5. Build a **park-bench** with the five or six straightest pieces on the east side of the **rotten-wall**.
6. Place the fire in front of your **thermometer-knees** when you are lying down, to monitor heat and minimize smoke problems.
7. Build an adjustable **vertical-log-windbreak** on the north side of the shelter to further control smoke.
8. Lean poles on the **rotten-wall** to create the structure for a **metre-square-sponge-roof**.
9. Obtain water from snow by melting it with the **slush-gutter** technique.
10. Protect your eyes with the **windshield-wiper** trick as you probe in the dark with a **wrapped-birchbark-torch**.

And so on.

This eventually became the framework for the course content of what I liked to call Realistic Wilderness Survival. No bullshit. Straight-to-the-point techniques to save your life while waiting for rescue. I taught this course to hundreds of forest workers all over Quebec, from Hydro-Québec field electricians to Ministry of Natural Resources biologists to bush pilots. I felt I was making a worthwhile contribution, perhaps even saving lives. Then I trained some of my best students into instructors to multiply the impact.

These multi-day courses are usually held after the first snowfalls to prevent forest fires. After theory and demonstrations, students are asked to surmount progressively more difficult simulated survival tasks, such as starting a fire with soaked materials. On the last day, they are invited to pretend getting lost in the wilderness just before darkness, to spend the night alone, with nothing but the clothes on their back and paper matches. We ensured their security with hourly whistle contact.

Of course, each of these occasions permitted me to discover still more of nature's wonders, since I would never miss the opportunity to experiment alongside the students. One time I chose to land in a very young forest in which poles of good dimension to make my park-bench bed were simply not part of the inventory. I resigned myself to make do with a score of maple saplings the size of my thumb. They rolled all over the place, so I used my shoelaces to intertwine each sapling with the next. Collecting and tying these was a pain, and all of a sudden I was weary of this suffering in the name of science. I wanted to change my vocation and leave further experimentation to the young and crazy. And then, as if Mother Nature had read my thoughts and wanted me to change my mind, I fell onto a new discovery. As I lay on the saplings, they bent under my weight, conforming to the shape of my hips, just like a soft mattress. What a luxurious feeling! Once again, I felt privileged to be a lowly apprentice.

And so I pursued my quest for survival knowledge and in similar ways developed many original techniques and wilderness survival strategies, which I published in a book in French whose title loosely translates to *Beyond the Survivathon*. These techniques I baptized with fun names such as: Lazy-Walk, Widow-Maker, Log-Hauler, Ember-Platform, Coffin-Wood, Thermic-Sponge, Zip-Match, Hook-a-Dee-Doo, 3D-Specs, Snow-Candle, Christmas-Tree-Signal, Da-D-Cup, and so on.

One of my favourite discoveries happened by chance much later. In my early fifties, although I appreciated comfort, I still enjoyed the occasional soft mattress between a fire and a rotten-wall, deep in the woods on a nice evening to get away from it all. But my pet peeve was the inevitable stiff neck from sleeping without a pillow. I had tried many strategies: itchy grass bundles, tickly evergreen branches, messy cattail fluff. Finally I had opted to do without. But one late afternoon I went out when I already had a kink. As I lay down on my park-bench, my neck hurt so much that I fumbled in the dark to find a log to support my head. I fell at random upon an old broken birch log with the bark still intact. But as I lay my head on it, the rotten wood inside the bark collapsed and dribbled out, leaving me with a twenty centimetre in diameter bark tube. I nonchalantly stuffed it with leaves, and out of nowhere, lo-and-behold, appeared the soft comfort of home. Nature would never cease to amaze me with its infinite tricks. The birchbark-pillow was the most surprising yet.

But not as surprising as the survival technique only my brother Michel could have dreamed up. We were younger then.

"Hey guys, why don't we go winter camping on the weekend, say Saturday morning to Sunday afternoon?"

"Yeah, let's. I'm sure we'll have a good laugh at Michel's expense again!"

The gang nodded in approval, everyone giggling except Michel, who was grimacing so gracefully that it changed the giggles into full-out laughter.

"But I don't have any gear!"

"Ah, come on Michel. You're the tough guy, you'll get by!"

"Okay, okay."

We were taking a break in the GB warehouse after unloading five trucks in a row. My brothers and a few of our regular work pals watched as Michel won hands down the turn-the-bottle-caps-inside-out competition. He's got tough fingers, that's for sure. He also won the carry-as-many-full-milk-crates-from-truck-to-fridge contest, and the stuff-as-many-cookies-in-your-mouth-as-possible challenge too. We went back to our respective jobs, wondering what kind of fun the weekend would bring. We agreed to invite other camping friends along.

After lunch on Saturday Michel arrived late, screeching his Batmobile to a halt with an emergency-brake slide right in front of us. To our great amusement, he pulled out a large old-fashioned flowery suitcase and tossed it on top of the other packsacks in the van.

"Where are your snowshoes Michel?"

"Bah, snowshoes are for wimps!"

"Michel will make himself snowshoes from his suitcase cover, just wait and see!"

We joked during the entire ride until we got to the boondocks and unloaded. After eight of us broke trail on snowshoes, Michel followed behind in our tracks. He had shoved a stick through the handle of his suitcase and carried it over his shoulder, hobo style. He's hilarious. We didn't hike too far before finding a suitable campsite, and Michel sure looked happy to drop his load. Most of us were too, for that matter, because we were grossly over-packed.

We set up our tents as Michel sat on his suitcase and relaxed, his back against a tree.

"Aren't you setting up your tent Michel?"

"Nah, I'll just sleep in that cave over there. Looks dry."

If he was referring to the rock overhang we passed by, he was in for a surprise. I saw signs of porcupines, and that depression was surely filled with scat. I started digging a trench and hole for the fire while others fetched firewood and logs to sit on. All the while Michel sat there joking, unwilling to move through the deep snow without snowshoes. We settled into a beautiful and windless March evening, and I pulled out the pots and pans to start our supper. People were fiddling in their packsacks and bringing out food to their chosen sitting spots around the toasty hot fire.

"What are you going to eat Michel?"

"I'll show you!"

Everyone was interested as Michel dragged his suitcase near to the fire and cracked open the lid. Everyone stared in disbelief. The only thing in there was a case of beer.

"Anyone want one?"

"Sure."

"Me too."

"Make that three."

"Okay, I'll trade you for a steak. And you some potatoes. And that spare blanket of yours."

I couldn't believe I too was succumbing and trading my extra parka for a beer. As gear fell on Michel's lap, it became obvious his survival strategy would work. Hats off to you, dear brother.

Two years later, I was in panic mode a couple of hours before departure for a trip with my students. I was running way behind schedule, having been way too busy finishing a research report. No time to pack food. So I headed to the grocery store and picked up a case of Pop Tarts. Only ate one. Feasted like a king all week.

Twenty-Three

The Opera Singer

"If you keep your emotions locked in a box, then when you want to open it one day you'll find that they're gone."
— James Thurber

After the memorable fortieth birthday party where my snow-filled canoe served as a beer cooler, wilderness adventures came to be few and far between. Not that there weren't many terrific outdoor trips. It's just that I was applying the risk management principles I was professing to prevent the occurrence of nasty events. But I managed to find my thrills in other ways.

Certainly work was fulfilling, as I pursued former interests with renewed vigour. With a colleague from the anthropology department, I founded the University of Quebec's first research laboratory in primitive technology. The advent of Internet permitted me to create the Primitive-Skills-Group so I could share findings with like-minded folk. I also enjoyed spontaneous experimentation, such as figuring out how Pierre St-Germain of the Franklin Expedition had managed to build a shrub-framed canoe out of a tent to cross the Coppermine River, exploring fire-by-friction in the physics lab using thermo-couples and infrared cameras, or inventing better emergency snowshoes when limited to using only shoelaces.

Similarly, I pursued my tripping food research, spending an entire sabbatical leave on this subject. I improved my trip food planning software until it was ready for marketing. But as everyone knows, I'm no good at that. So I made it available for free and it is now used all over Quebec.

On the environmental side, I became a Leave No Trace Canada master educator instructor to influence the movement from the inside. I was mostly successful at this, having helped rewrite the Canadian reference booklet and training those new instructors that would go on to offer the master educator courses in Quebec.

But certainly the highlight of my career was founding the first ever Bachelor's Degree in Outdoor Adventure and Tourism with my long-time buddy and colleague Mario Bilodeau. Together, we designed a unique program with two intensive sessions taught almost entirely outdoors, whose main purpose was to form leaders for Quebec's expanding backcountry tourism industry. Like dedicated parents, we took care of our baby by creating all sorts of stimuli. For example, to teach security with tools we had vans drop loads of full-length logs on campus so students could practise, which also permitted them to earn money once they had sawn and chopped them into firewood. We invited the rector and other dignitaries to thematic banquets students organized in the outdoor classroom we had built in the woods behind the campus. We designed major first-aid and rescue simulations that felt real. We challenged them to beat the existing record for the knot board I had designed.

I directed the program for the first six years, fighting to obtain funds, technical staff, and dedicated spaces for gear. I also founded the Outdoor Research and Expertise Laboratory (LERPA) as a support structure. That took a lot of steam out of my locomotive. But the results were worthwhile, as we watched our offspring inundate Quebec with innovative ideas.

Upon conception, based on the success of the Changing Gears project and also a Mount McKinley climb by Mario and his apprentices, we imagined that one of the hits of our program would be to add a requirement that all third-year students organize a major expedition. But we hadn't realized to what extent our over-achiever clientele would go to push their wild projects to the edge of acceptable adventure; I don't know why we didn't grasp that they were just like we were at their age. All of

The Opera Singer

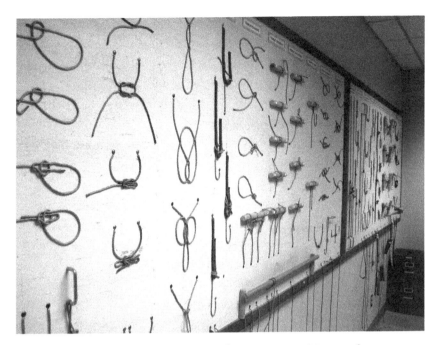

The university's knot board. The record is 5 minutes, 38 seconds.

a sudden, Mario and I were overseeing cross-Canada or Baja California kayak expeditions, a ski journey traversing Ellesmere Island, climbs to the summits of Mounts Denali and Logan, treks around the Himalayas, paddles on whitewater rivers in the Arctic and remote Ecuador, cycling tours from Canada to Patagonia, a month-long survival experiment, and therapeutic adventure expeditions with youths faced with cancer, to name just a few. This forced me to write up quite extensive risk-management procedures that were eventually published conjointly by the Quebec Camping Association and Adventure Ecotourism Quebec. All of which took time. I felt like I was a prisoner chained to my desk, allowed but a brief fitness break daily while I watched everyone else roam free.

As countless students returned from expeditions to relate events with eyes bright as brand new coins shining in the sun, I identified the common denominator defining adventure — pure and powerful emotions. Nature has a way of intensifying things. Whether flabbergasted by an exciting discovery, scared to death that an ice bridge would collapse, happily overcoming a seemingly impossible obstacle, tenderly facing lost friendships at the end of a trip, being awestruck by amazing scenery, spitting anger at the injustice of spoiling a wild river with a dam or garbage, witnessing cruel inter-animal attacks, or suffering nauseating cramps and agonizing blisters, the emotions in the wilds are so heartfelt they make tears flow. They prove you're alive.

I longed once again for those expedition-borne emotions. Breaking my spine by ramming into the boards during a hockey game didn't help. For months I had to wear an awful corset that made me aware for the first time that I was a mere mortal. Although I fully recovered from the bad luck, it fattened me by five kilos. And as a chef, my adoration of rich foods prevented me from ever again enjoying a six-pack silhouette. At least the accident had provided a new sensation.

Someone once stated that you know you're getting old when you'd rather talk about the things you've done instead of the things you're going to do next. When is the last time you did something for the first time? It seems to me that adventure-type emotions only occur when we opt for a touch of newness.

One project I had had in mind for a while was constructing my own birchbark canoe. It took over two weeks to chase down materials in the forest. Then I forged the crooked knife and the awls I would need and built a traditional shave bench, for I wouldn't allow myself to cheat by using modern tools. I also found a froo, a tool used for splitting board, for which I made a handle, and cut a metal barrel in half as a soaking/boiling tub. After splitting and whittling cedar ribs, planks, and gunnels until I was buried in shavings, I built a wooden platform in my barn on which to build the canoe. (Years earlier I had purchased an 1853 schoolhouse with a barn and had renovated the first into a unique small home while the second served as a spacious workshop.) I sat many hours sewing with tamarack roots, then installing the planks and steam bending the ribs to force them into place. I also hand sewed an Egyptian cotton sail, because I wanted a sailing rig like the one I had seen in Adney and Chapelle's reference book.

After a total of nine hundred hours of toil the canoe was ready. I portaged it to the Saguenay River and must admit to tears of emotion as

Sewing a birchbark canoe.

I christened it and paddled away. After a few exhibitions, I legged it the museum of the Centre d'Interpretation de la Métabetchouan, where it still sits proudly.

Surely one of the most intense sensations of my life occurred after I had just obtained my airplane pilot's licence. Ecstatic is a small word to describe my sentiment during that first solo flight, especially when illegally popping targets of miniature cloud puffs at ten thousand feet. The hysteric joy made me yell like a madman. But flying an airplane also resulted in the three downright scariest moments of my life. During my first flight from Quebec to Ontario I was soaring way up there, admiring the spectacular view of the Thousand Islands. The visibility was infinite; a gorgeous day. I synchronized the weather channel, to respect procedure. It said that Toronto was presently enclosed in a storm system with less than thousand-foot ceilings. Like a lost person not believing his compass, I dismissed the weather warning as an old recording. But an hour later I saw it. A black wall. In all my years as an outdoorsman, I had never realized that weather changes were cut so square. I marvelled at this awesome back seat-view of nature for a bit. But then I wondered what to do. Too inexperienced to calculate an alternate landing spot on the fly, I just pushed on, drawn forward by the anticipation of arrival and the complicated ground-transport logistics of a detour.

While pilot training in Chicoutimi, the multitude of uniquely shaped lakes had become easy orienteering reference points. But as I flew into what seemed to be the eye of a hurricane just outside Toronto, the infinite criss-cross of roads prevented me from getting my bearings. I was in way over my head. The turbulence repeatedly seemed to yank the jerking airplane out from under me. With sweaty palms I held on like a granny on a too-wild amusement ride, blindly following my bearing toward Buttonville airport.

Even with the help of the control tower, I couldn't distinguish the landing strip from the thousands of streets below. I flew through the downpour and right over the airport. After two loops and the tower's patient guidance, I finally spotted the runway. However, during flight training I had never landed in over seven knots of crosswinds. And now, in the middle of this mess, they were gusting at thirty-five. I landed with a ten-metre bounce, pitching and wavering so badly the right wing just barely missed the ground. Dad had come to pick me up and met me at the gate, wondering why I was jumping in his arms with such fervour, white as a ghost.

On another occasion I landed on a too-short private airstrip in Huntsville. I had no way of knowing the runway was slick with ice, and slid right off the end and up a hill until I ran into a rope fence that wrapped around the propeller. Thankfully there was no damage, and so once again I got off with a good scare — hopefully the last. But the very next day as I was flying out of that hole, the melting ice slowed my takeoff and I barely made it up over the trees; I swear I heard the wheels scratched by branches. To make matters worse, when I got to Ottawa the fog had not lifted as per the weather report, and I resorted to flying just above the river valley underneath a ceiling of but a few hundred feet. I asked Ottawa's tower for help, and they suggested I switch my destination to Rockcliffe Airport on river right to avoid flying so low over the city. Then they warned me about the high bridges. What high bridges? Yikes! THOSE high bridges. I could barely squeeze in the narrow passage between the trucks and the clouds. There was Parliament Hill, at eye level to my right. I was sure glad to see that runway. Stress caused me to land way too fast; it wasn't a pretty sight. I parked the aircraft, swore every nasty word I knew, and hopped on a bus for a nine-hour ride back to Chicoutimi.

It turns out I had delayed the landing of a few 737 jets at the Ottawa's Macdonald-Cartier International Airport and was summoned to file a report to the Ministry of Transports to explain why I was flying in those nasty conditions. The next day I paid two pilots to go fetch the rental plane. Although I obtained pardon from the Ministry, I did give up flying, preferring to stay alive. Whew!

I thought I had felt every emotion. But no. At forty-three years old, I stormed into the university staff room to pick up my mail and stopped dead in my tracks. There she was, of rare beauty, glowing while cutting little blocks of cheese. "What are you doing with that cheese?" I asked.

She gazed at me with mysterious eyes, smiling. "Want a piece?" I liked that evasive answer, so I popped one in my mouth.

"Yech!" I exclaimed without thinking.

"Oh, it's non-fat cheese I pass around during my conference on nutrition. It's a substitute when you want to lose weight."

By then my head was spinning, I was at a loss for words. So I exageratingly stuck my stomach way out and blurted: "Well look at me, I'm way too fat. Maybe I should come and listen to your talk!"

Those must have been the words Lizon wanted to hear, because three years later, Véronica was born. As every parent knows, nothing in the world can compare to the ecstasy of watching this true love grow. Her first canoe trip at two weeks. Her first winter camping trip at three months. Her first fish at a year and a half. I was in heaven. From two years old onward, she loved dressing up to participate in the annual traditional activities colloquium I organized with my students. Probably because she was the hit of the "village." The apple doesn't fall far from the tree. Fortunately, my lady partner's serenity counter-balanced my extravagance.

As Véronica grew taller, I decided to build our family a wider nest. The massive barn window offered a spectacular view of the Saguenay River and the Valin Mountains beyond. The tiny farmhouse window offered a spectacular view of the barn wall. That was about to change. I would retire in a home with a lovely panorama. And with soul, I hoped. But building-related choices are not easy with an environmentally conscious mind. Especially an extravagant one.

It took seven years. I cut trees by thinning them here and there on my land. Had them sawn with a portable mill into lumber. Tore down the barn and kept all the best beams and greyed-out planks to recycle them into the new house's decor. Piled up house parts purchased at auctions and second-hand places. Found a plan on the net and modified it myself. Hired a construction coach and a couple of neighbours. Put on the carpenter's apron.

Véronica in the traditional village.

The Opera Singer

I long considered alternative options, like geodesic domes, straw-bale houses and such, but chickened out. Also built the house too big, probably, and with too many windows for proper heat retention. But it's nice. The central masonry contra-flow heater uses less firewood, which compensates. And its enclosed bake oven heats for free. Every corner of the house holds a memory. Like the baking counter built from the wood of the tree whose bark I used to build the canoe. Or Véronica's rocking horse. Or the ten-metre-long beam holding up the vaulted ceiling. This most extravagant of survival shelters emits feelings of pride. As for the old schoolhouse, moving it to a new location down the road preserved the heritage.

From the living room rocking chair retirement age appeared on the horizon. Maybe it was time to activate my sailing dream and relax. After purchasing a Siren 17 pocket cruiser, lugging it back from Ontario in February, and spending winter fixing it up, I was ready for my first sail. The shop had totally overhauled the motor at an exorbitant price, so I felt confident in this backup option. Tadoussac's music festival was a good occasion to take excited Véronica with me for a sleepover at the town's marina; that way we would avoid the overcrowded main parking lots and try out our boat at the same time.

Tadoussac marina's boat ramp can only be used at high tide. Oops. Not wanting to wait four hours, we decided to take the car ferry across the mouth of the Saguenay River, admiring the majestic St. Lawrence River, gateway to the sea, and launched the boat from the cement ramp there. Then we motored back across the calm waters, and I managed to dock the sailboat into the marina slip with just one smash. After lunch the winds rose, and we went out into the bay for a first ever attempt at sailing. I prudently hoisted the main sail only, double-reefed, and all went well, except for a couple of unexpected jibes of course. Satisfied, we docked again and walked to town to enjoy the festival's invigorating music. At the end we were pleased to leave the crowd and return with anticipation to our miniature floating cabin.

The next day brought blustery winds and it was obvious we could not cross the channel back to the van. I chatted with old captains who advised that the traverse would not be possible until late afternoon. So we waited. And waited. Finally the locals confirmed it safe enough, although the waves were a bit tall to my liking. I wouldn't sail, just motor across. We donned our life vests and safety harnesses and soon Véronica was really enjoying the wild roller-coaster ride. I wasn't. But we were already two-thirds of the way across, only a few minutes from the cliffs on the opposite shore. That's when the motor quit. No panic yet. I hoisted the main sail and got back under way. But two minutes later the wind completely died, leaving us bobbing like a cork on crazy leftover waves. The cliffs were too close for comfort, so I pulled out the paddle and soon was sweating profusely, to Véronica's shouts of encouragement. Brave girl! Turns out one of the captains had been watching all this through binoculars and had advised the coast guard to come to our rescue. I hadn't even had the presence of mind to call them myself! Parenthood sure magnifies sentiments.

I reflected on my journey thus far and the gamut of emotions it had provided. In the wilderness, from excitement and wonder to worry and fright. At home, from utter joy when my daughter dives into my arms to the excruciating sadness when my best friend Dad disappeared. All part of nature in the larger sense. But there was one emotion I had not yet experienced.

"Welcome to our music school. What can I help you with?"

"Hi, I'd like to sign up for a singing course."

"Okay. What kind of singing?"

"You know, the kind where you use your voice."

"Ha, ha, very funny. I mean would you like to try popular singing or classical singing."

"Huh, popular singing I suppose. But what's the difference?"

"Well, popular singing is more fun, but if you really want to learn to control your voice you should go for classical. Many of the top popular singers have a classical singing background."

"Who's the teacher? Is he good?"

"Réal Toupin. He directs the symphonic orchestra choir. We're lucky to have him; he comes up from Quebec just once a week to help us out. He's toured all over the place. Unbelievable singer."

"Sign me up!"

The Opera Singer

"Would you like theory too? It's free. Do you have any experience in music?"

"I'm mostly self-taught, but I've taken a few guitar lessons. I've been playing for thirty years and I've sung in bars a bit, but mostly around campfires. Sure, I'll go for theory too."

That's me alright. Always choosing the hard way. I went there just to improve my voice a little, and now I was signed up for the whole package. And knowing me I wouldn't quit until I succeeded.

Music has haunted me all my life. That's one thing I had no talent in. Whereas my teenage pals would pick up an instrument and play after a couple of months, I still couldn't tune my guitar after five years. And twenty years later, when I purchased a keyboard with included rhythmic patterns, it became obvious that I couldn't keep a beat either. No wonder no one wanted to jam with me!

But I wanted to play so badly that I kept time by watching the little blue light flash on the machine's dash until I felt the associated beat in my body. It took months, years even, but music finally penetrated my thick skull. I deserved a medal for tenacity. I'd even been writing songs, and would like to record them for my friends and family. But first I'd take a few singing lessons, just to make sure my voice was okay.

Wednesday night at seven on the dot, a student walked out the studio door and it was my turn to meet the famous Réal Toupin. His bright piercing eyes and huge smile welcome me as I shake his hand. Not five seconds elapse before we're working.

"Sing with me. DO. Do mi sol do sol mi do. RE. Re fa la re la fa re. MI. Mi sol si mi si sol mi. FA. Fa la do fa do la fa. SOL. Sol si re sol re si sol."

On the last re I cracked. He informed me that I'm a tenor. To sing higher, I would have to raise the roof of my mouth to expand the cavity, as when yawning. He made me yawn ten times.

"Okay, again. DO. Do mi sol do sol mi do. That's better. Keep going."

We vocalized like this for a whole hour, non-stop. This guy had so much passion it was unreal. By the end of the course I'd managed to emit my first fa.

Week after week, I religiously returned to my singing class. It's like Latin. Discipline, and lots of it. But Réal was so darn intense he made it fun. After five months, I managed a G note. Now for the A flat. There's a passage between G and A flat apparently. Makes it a money note. Which means you can earn more money when you can sing it.

We started practising my first song. It was a French classic, "Je t'ai donné mon Coeur," which means "I gave you my heart." More poetic in French somehow. Réal had chosen it for me because it ended on a sustained A flat. My first attempt was disastrous. Not only could I not reach the high note, I couldn't vibe the triplet rhythm. This motivated me to practise more at home, or in the wilderness when I wanted to chase the bears away. Another three months and I could finally sing the song, albeit with difficulty.

"Maybe you should try singing it at the end-of-year show?"

"Sure, no problem."

I hadn't noticed the word *try* in his sentence. Bad mistake.

It was the end of April. I'm standing under my umbrella in the woodlot outside the Mont-Jacob concert hall, vocalizing. Beyond hearing distance, I see limousines drive up. Tie-clad men and chic well-dressed ladies enter the theatre. Lots of them. Better go inside and get ready. I passed by the poster announcing the event, the same one that was in the weekend paper. Inside, the decorum looked as it should if the queen was in town.

Backstage I recognized many of Réal's long-time students I met at a master's class he had organized, where he had invited his own teacher, diva Jacqueline Martel-Cistellini. Every single one was wearing a fancy tuxedo or a long gown. Except me, in white shirt and suspenders. Réal informed me that I would be going first. What?

I started getting nervous. I fled outside for a minute and vocalized a bit. Stress. I went back to find comfort in my teacher, but he was too busy coaching his stars. Some of the others were peeking out at the crowd between a crack in the curtains. I dared not.

The master of ceremonies was greeted by the crowd with thunderous applause, which doubled in intensity when he presented the professional musician who sat at the grand piano. As he presented my name, I wondered if the audience associated it with the survival-expert me. I got a tap on the back from my coach — or was that a shove? I walked out on stage and heard stifled laughter. Then, all of a sudden, it was quiet, oh so deadly quiet. My throat was dry. Help! I gave a faint smile to the pianist as I took position. There were three hundred pairs of eyes peering straight at my face. More specifically, focused on my lips. I freaked. My legs were literally shaking in my pants; I could not physically control them.

The piano strings vibrated to project the beautiful melody into the audience. There was my chance at fame. Breathe André, try to relax. But I thought I was going to faint instead. My first notes seemed okay. Hey, not bad in fact. After the first verse, I'd gained confidence. I was going to make it through this. My voice was powerful, said Réal. Concentrate, think of yawning. Here comes the A flat. I belched it out. Squawk! The most nauseating, horrible, ugly-duckling squawk anyone has ever heard.

They applauded out of pity. I exited with head hanging in shame. Pure shame. Unadulterated shame. Absolute shame. Ouch!

Conclusion

"The worst risk of all is to not risk at all."
— André-François Bourbeau

Réal Toupin patiently helped me regain my confidence. But it took five years before I dared to sing in public again, this time as member of the symphonic choir. I eventually learned that music has very little to do with performance. It is a tender gift you give friends. It seemed to me worthwhile daily occupations should be like music. At long last, I could abandon the search for recognition through exploits.

A while back I was invited to sing the Panis Angelicus at a service in Champlain. I said yes because the 1671 church there happens to be a National Historic Monument and it was built to naturally amplify unplugged voices. Plus, the incredible original organ still tosses notes way up to the vaulted ceiling. My rendition was far from being perfect, maybe not even good, but I gave it my soul to honour history. People smiled. Perhaps because I had finally become a musician. Or perhaps because the tuxedo I was wearing looked nice. No matter, it was a fantastic experience to time travel back to Champlain's era.

It is certainly peculiar, this strange bird called Time, flying by so quickly, yet still managing to gently deposit the basket of maturity on our doorstep. With age and experience, we see differently, weigh things more accurately. But frustrating too, this gift of newborn wisdom, because we realize we only experimented with but a tiny fraction of the dreams we caressed while young. Alas! I came to realize that in a single life I could only scratch the surface of all the secrets nature had to offer. Good reason to pursue the crusade for wilderness protection.

There are occasions in life that remind us of our age. Like the first time a student called me Mister Bourbeau instead of André. Or worse still, the day one of my pretty twenty-year-old students exclaimed "Oh, Dr. Bourbeau, it's so much fun being in the bush with you. You're a bit like our grandpa you know." Yikes!

Well, I decided to push the grandpa idea to celebrate the last course of my university teaching career — traditional outdoor activities. I had opted for a gradual retirement plan, thus I had some time. I wanted to give my students and myself a final splashy special project, so I imagined a scenario where we were a bunch of farmers from Laterrière village who had been hired by William Price, Saguenay's eighteenth-century timber baron, to go fetch two immense wooden beams for the construction of a church. Each of us had a name and a role in history, from schoolteacher to physician. I was the old granddad of the gang, a trapper.

I put the students in the mood by visiting Chicoutimi's old pulp-mill museum, giving them copies of the newspapers of the era and circulating copies of the original 1895 Montgomery Ward & Co. catalogue. They would prepare their own costumes and personal gear. Oh, and the guys had to grow a moustache. They would also research their character, plus find a subject to teach the others. My assistants and I would take care of everything else: food, tools, group gear. We enjoyed eight months of adventure begging for all the old stuff we needed, or making it ourselves, or buying it. It cost me a fortune, both in time and money.

But what a neat voyage to the past we lived that November. Unknown to the students, I had convinced a friend to come with a giant draft horse and sled to carry the gear to the end of the bush trail. From there it was a bushwhacking portage to the camping site in a soon-to-be-cut forest, so we could freely chop all the wood we wanted. My assistant had planted a farm-raised deer in the forest and covered it with fake blood so he could

Conclusion

go "hunting." When we all heard the shot, I sent four students to go help him drag back the meat we would eat for ten days.

At first the men lived in a canvas lean-to. But after five days of axemanship, they were installed in a typical lumberjack camp, complete with central fireplace. The gals lived in a prospector's tent, owned by the guy who had been to the gold rush, while the bosses had their own baker's tent. As per tradition, we prayed every morning and every night, and said grace before each meal while gathered around the outdoor Chippewa kitchen. On Sunday we served mass in Latin, thanks to a colleague who donned the black robe and came to visit, with a choir to boot.

We felled a giant white spruce and squared it with broad axes, holding it steady with iron dogs made on the spot by the blacksmith of the group. Then we pulled it half a kilometre to the lake, using hemp ropes and wooden pulleys. For a final highlight, I had my old Retro-Propulsion-paddler Marcel disguise himself as Sir William Price himself. He was a terrific actor, English accent and all, as were his wife and son. By then we were so totally immersed in our roles that his arrival gave us

Granddad with Sir William Price and family, and Father Tremblay.

all goosebumps, even me who was in on the secret. That night we partook in homemade bread, baked beans, roast deer ribs, and apple crisp, followed by a rousing sing-along led by the resident teller of legends. A wonderful end to a wonderful career.

When I retired, I was honoured with Professor Emeritus status, which means I'm allowed to continue working for free. Joking aside, it mostly signifies that I can continue my research while accessing the university's resources. I suppose it also gives me the right and obligation to write wise conclusions in books. I'll try. In fact, I'll abuse my privilege and offer three.

For my first conclusion, I would like to discourse about risk.

I remain convinced that risk is worthwhile. The worst risk of all is to not risk at all. Risk defines adventure, and adventure defines life. To feel alive, one needs to push limits. To fall down sometimes.

But that doesn't mean taking risks at all costs. There are good risks, but also stupid risks. Good risks are the ones whose result, if the danger materializes, make you suffer from blisters, bring you to the edge

The joy of walking on thin ice.

Conclusion

of exhaustion, make you wet and miserable, make you scratchy from mosquito bites, scare you to death while staying safe, cost you money by scrapping your gear, make you look ridiculous, or even shame you to tears. Stupid risks are the ones whose materialization kill you or maim you permanently, like losing feet to frostbite. I've taken a lot of good risks in my life, which made me grow stronger and made me a better person. I also flirted with many stupid risks, of which I am not proud — these I was simply lucky to survive. Those types of risks I will not take anymore. Nor would I suggest them to anyone. Life is just too sweet to waste it by accident.

Yes but, some will say. I know, I know, we risk dying every day in traffic. It's the old risk-versus-probability graph. That doesn't mean we should play ball in the middle of the freeway. Consider the historic difference in canoeing risk. After spending hundreds of hours building a birchbark canoe, the coureurs de bois thrilled in descending class 2 rapids, the consequence of which was having to start all over again after walking fifty kilometres back to camp. Good risk. Nowadays, with ABS canoes, that thrill is gone, because the boats are indestructible. So we move up to R4 and R5 rapids, where the chances of drowning are severe at the slightest mistake. Bad risk.

To help manage risks, I have found that it helps to consider four simple pessimistic scenarios before undertaking an adventure. What happens if I'm detained for a few hours for various reasons? In what kind of trouble will I find myself if I lose my gear, for example if my motor quits on me? Where will I get help if I'm badly hurt at the furthest point of my route? What are the consequences of a human mistake? If the answer is "I won't like it cause I'll suffer," you may choose to proceed. But to the answer "I die," I would strongly reconsider.

For my second conclusion, the discourse relates to wilderness survival.

Once we're in trouble, my observations are that survival depends on four factors: physical capacity, psychological outlook, specific technical skills, and decision-making. And let us be aware that my own experiences have all been piddly compared to real survival situations. After all, I was doing this stuff voluntarily, and almost always had a way out. I was just playing. When we study real incidents, when life is truly at stake, survival takes on a whole different meaning. When bones are broken or guts are hanging out, we're no longer talking macho. Survival is no fun then.

Of the four factors, physical capacity can be improved by increasing fitness level. My mantra on this one is simple: GO PLAY OUTSIDE. As far as psychological outlook is concerned, you have to love life. DO THINGS. It also helps to read real-life survival stories. When you do, it gives you examples of what others have been through, and if it happens to you, you have a reference point. Technical skills can be acquired through the long and narrow path of experience, but they are faster and more safely gained by engaging in simulations. TRY CHALLENGES. Enter the loop of try-fail-try-fail-try-fail-try-WIN.

But although these three factors are all fundamentally important, it's often the decision-making that kills or saves. The study of close-call incidents and fatal accidents reveal this clearly. Over the years, with the precious help of my colleagues at the university, I have developed a tool for thinking straight when caught in a wilderness survival predicament: the SERA model.

SERA is an acronym for Search-and-rescue, Energy, Risk, and Assets. The key to survival is to always stop and consider the impact your decision will have on each of these four factors before action. THINK SERA BEFORE ACTION. Qui sera sera. Whoever will be (thinking sera) will be (sera) alive.

KEY POINTS FOR SURVIVAL IN THE WILDERNESS:

S – Help **Search**-and-rescue teams to find you
E – Conserve your precious vital **Energy**
R – Minimize **Risks**
A – Pamper your **Assets**

In concrete terms, this translates to preparing signals and staying in open spaces, immediately switching to slow-down-and-save-energy mode, avoiding all further perilous acts, and taking precious care not to lose or otherwise destroy or damage the gear available.

Conclusion

The SERA model's role is to prevent foolish immediate action (panic), which is perfectly natural in a stress context (let's get the hell out of here or let's do something about this right now).

For example, your snowmobile breaks down, it's ten below zero Celsius with wind, early evening, forty kilometres from camp. One option is to build a Quincy snow house. Stop. Think SERA. S. Inside a Quincy, I will not hear the search-and-rescue snowmobile go by. Negative. E. It takes a lot of energy to build such a shelter, but I will save energy fighting cold once built because I will be out of the wind. Neutral. R. If the Quincy caves in on me while building it, I risk suffocation. Negative. A. Right now my clothing is dry. Once the Quincy is finished, they will be at least partly wet, both from snow and from sweat. Very negative. Hmm, maybe a Quincy isn't such a good idea after all.

Another option is to start walking out along the snowmobile path. Stop. Think SERA. S. Positive. E. Very negative. R. Very negative, hypothermia and frostbite if I don't make it. A. Negative, clothing will get damp. Better forget that idea too. So I just slowly dig a hole behind my machine to get out of the wind and wait, doing calisthenics as need be to maintain slight warmth.

My third and final conclusion does not refer to wilderness survival but rather to survival of the wilderness.

It was gently snowing last October as I was standing on the shore of the heart-shaped lake where the helicopter had dropped us for the Survivathon. Yes, roads now permitted me to drive my van way up into this formerly wild territory. But I couldn't recognize the spot where I had lit my magical fire a quarter of a century before; the forest was no longer, it had fallen to clear-cut logging. Although I understand the economics of forestry, which can be viewed as agriculture spanning sixty years, and know full well that forest eventually rises again, I could not help but feel a certain pinch faced with the desolation before me. This made me question the qualifiers "progress" and "development" to justify human gluttony.

I am often asked if I would repeat my Survivathon or other adventures. Of course not. I'm not a masochist. But if I could go back in time and was permitted to choose again, I would not hesitate a single micro-instant. Because there are no words to describe the raw power of this intimacy with nature. It is like watching a child grow. What emotions! And just like a mother forgets the pain of childbirth while she later admires the eyes of her growing baby, I have long forgotten the labours I lived through and only my soft regards for tender nature remain.

The value of a passionate relationship with nature cannot be expressed, just like a picture can never bear witness to the smell of flowers, the sound of a creek, or the feel of the wind on our faces. We have to be there. And when we do sense this great Cathedral of beauty which combines both complexity and simplicity, peace seeps into our pores. And we care.

This is why I invite young and old to play "wilderness survival." Each fire, each tasted cattail, each sip of water drunk from a bark cup, and each wiggling trout on the hook adds perspective to our soul — necessary in a modern life that artificially chops us off more and more from our origins.

In my years of teaching outdoors, I've had the opportunity to observe many expressive faces: wonder at the appearance of a first ember by friction, satisfaction when lying on a successful park-bench, pride as the slingshot stone smacks its target, surprise when birchbark ladles mirror in water, sweet smiles when the log raft floats away in the mist. Then, several years later, it is with contentment that I admire these same faces — aged a bit, mind you — in positions of leadership in a variety of roles in society, but always imbued with a common interest in protecting nature.

It is probably my mentor, the late Kirk Wipper, who best expressed my thoughts. "You have to do what you can, do your best with what you are. And you have to believe in wilderness. If you do that you can't go wrong." So, as a tribute to Kirk, I would like to raise a point that I believe is fundamental to the survival of the wilderness. I call it the construction disease: the uncontrollable mania of the human being to build and develop, at any expense, justified by all kinds of pretexts all allegedly greater one than the other. Thus, we see the proliferation of new forest roads, trails, hunting vantage points, cabins, chalets, lodges, dams, and so on. Armed with chainsaws and more and more efficient all-terrain vehicles, forest tourists mow their way to each lake or pond until they can access it by motorized means. Once the trail is constructed, they install basic comforts. But in no time they require fridges, stoves, and

Conclusion

then television. In short, they have moved the entire city to the forest. Thusly, nature and wild spaces gradually disappear. The construction disease fiercely attacks — even in the supposedly protected territories called parks, customer demands force administrators to develop additional infrastructures.

The construction disease also infects cities. Saguenay has at least doubled its surface area in the last two decades, without increasing its population. Where we once saw nature with buildings here and there we now see buildings with nature here and there. In my immediate neighbourhood, the last thirty years has seen emerge a factory and countless houses built at the water's edge, until there is now no place to camp. Even the agricultural zoning did not resist the onslaught of developers who easily received the rights to construct.

I have nothing against the construction industry. I see no harm in replacing an obsolete building by a new one, in beautifying a neighbourhood, in re-asphalting a road. What I question is this tendency to solicit

Bannock phantoms.

new wilderness areas in the name of progress. This encroaches on nature, irreversibly destroys it. The result is that we are now surrounded by city, rather than by nature. No one can accurately predict the effect of this change on the quality of life of future generations. To eradicate the illness, I therefore vote for a moratorium on any new construction that impacts wilderness areas. One day, perhaps, annual development and progress will be measured by the number of buildings that have been removed to yield place to nature once more. I hope so.

What's next for me, I wonder? Last summer I paddled the entire length of the Horton River in the Northwest Territories all the way to the Arctic Ocean. Caribou, muskox, bears, big fish, good friends. I liked that. My mind is boiling with fun ideas. Hike the Pacific Crest and Appalachian trails. Sail the Great Loop. Canoe Mongolia's rivers. Mountain bike the Continental Divide. Build another spruce-bark canoe and paddle it as a coureur de bois through Cree territory. Not for performance. But to leisurely enjoy simple, non-motorized travel. I will also surely pursue my research work — after all, research is the infinite final frontier.

And if one day my health limits me to strolling in the park, playing chess, doing the daily crossword puzzle, reading and playing in a band, perhaps I will sit down again to write another book concerning my new adventures. But until then, I'm off to the wilderness. Cause I'm hungry for more bannock phantoms.

Favourite Books

These are my five favourite English-language books of all time. It wasn't an easy choice.

Lennard Bickel. *Shackleton's Forgotten Men: The Untold Tale of an Antarctic Tragedy*. Boston: Da Capo Press, 2000.

Richard Buckminster Fuller. *Critical Path*. New York: St. Martin's Press, 1982.

Steven Callahan. *Adrift: Seventy-Six Days Lost at Sea*. Boston: Houghton Mifflin Company, 1986.

A.J. Mackinnon. *The Unlikely Voyage of Jack De Crow: A Mirror Odyssey from North Wales to the Black Sea*. Ann Arbor, Michigan: Sheridan House, 2002.

Dillon Wallace. *The Lure of the Labrador Wild*. Grand Rapids, Michigan: Fleming H. Revell Company, 1905.

And here is a pretty good one in French. This choice was easy.

André-François Bourbeau. *Le Surviethon: 25 ans plus tard*. Chicoutimi, Quebec: Éditions JCL, 2011.

About the Author

Photo by Michel Martineau.

André-François Bourbeau was born in Quebec's Eastern Townships but was raised in Northern Ontario. After completing two bachelor degrees at the University of Toronto, he pursued post-graduate work at the University of Northern Colorado where he earned a Masters of Arts in Outdoor Education and a Doctorate degree in Wilderness Survival Education. He spent his career as professor of outdoor pursuits at the University of Quebec at Chicoutimi, where he founded Quebec's only research laboratory that studies wilderness survival, outdoor risk management and emergency response, therapeutic adventure, and expedition logistics. He has spoken at hundreds of conferences and authored numerous works in French, his most recent book being a treatise of his wilderness survival research results.

Upon retirement Dr Bourbeau was honoured with Professor Emeritus status, which permits him to pursue his research activities at his leisure. He currently lives with his spouse and teenage daughter in the Saguenay region, a wonderful place to enjoy spending time in the wild outdoors.

Professor Bourbeau shares more interesting wilderness secrets at *wildernesssecrets.com.*